A Mathematical Gift, I

The interplay between
topology, functions,
geometry, and algebra

Mathematical World

Volume 19

A Mathematical Gift, I

The interplay between topology, functions, geometry, and algebra

Kenji Ueno
Koji Shiga
Shigeyuki Morita

Translated by Eiko Tyler

AMS
AMERICAN MATHEMATICAL SOCIETY

KOKOSEI NI OKURU SUGAKU, 1

Kenji Ueno, Koji Shiga, Shigeyuki Morita

Copyright © 1995 by Kenji Ueno, Koji Shiga, Shigeyuki Morita

Originally published in Japanese in 1995 by Iwanami Shoten, Publishers, Tokyo, Japan

This English edition published by the American Mathematical Society, Providence, Rhode Island, by arrangement with the authors c/o Iwanami Shoten, Publishers, Tokyo

2000 *Mathematics Subject Classification*. Primary 00–01.

For additional information and updates on this book, visit
www.ams.org/bookpages/mawrld-19

Library of Congress Cataloging-in-Publication Data

Ueno, Kenji, 1945–
  [Kokosei ni okuru sugaku I. English]
  A mathematical gift I : the interplay between topology, functions, geometry, and algebra /
Kenji Ueno, Koji Shiga, Shigeyuki Morita ; translated by Eiko Tyler.
    p. cm. — (Mathematical world, ISSN 1055-9426 ; v. 19)
  ISBN 0-8218-3282-4 (alk. paper)
  1. Geometry. I. Shiga, Koji, 1930– II. Morita, S. (Shigeyuki), 1946– III. Title. IV. Series.

QA445.U3613 2003
516—dc22   2003062778

1519

Contents

Preface

About two years ago, we developed a series of lectures on mathematics for high school students, which were held every other week at Kyoto University. Kenji Ueno, who is one of the authors of this book, began this program and was primarily responsible for initiating these lectures. This unique lecture format took root and grew due to Ueno's keen understanding of the education and learning process and the support and enthusiasm of approximately fifty high school and junior high students. Today you can find this mathematical lecture format not only at Kyoto University, but at many other universities in Japan.

On December 23 and 24, 1994, Kyoto University held a special program of lectures on Geometry as part of this series. The three authors of this book gave their presentations on different themes. After the scheduled lectures, two of the high school students who had attended asked if we have any plans to publish a book about the material we had just covered.

As it turns out, we were thinking about publishing a book based on this material because we felt that the lecture format could not reach enough people. Once we limit the presentation strictly to lectures, the audience drastically narrowed. Also, geographical limitations further reduce the number of people who have access to the information. So when we were asked if we would publish, we were determined to do so. At that time, though, we were not yet prepared to arrange and publish the lectures.

One night, one of the students suggested "Mathematics as a Gift to High School Students" as the title of the book. Initially, this suggestion surprised us. Inasmuch as our lectures to high school students were the same as those given to college students, we did not consciously differentiate them. The high school students had, however, picked up on a point that both we and our college students had missed—that education is a gift. It is a legacy that should be handed down and talked about through generations. I believe there is a similarity between the concept of learning and the concept of land as something to be inherited from those who came before us: something that can be handed down from generation to generation as a precious and irreplaceable object; I think we have lost sight of a very important philosophy that education is an invaluable gift. Something clicked in our minds that night. We decided to publish a book with the title the student suggested. The decision, which was made on the evening of December 24th, was a suitable one for Christmas Eve.

With the assistance of Iwanami-Shoten, our publisher, we are able to present this two-volume gift of mathematics. It is our hope that our readers will find the contents of these volumes to be like the joyous sound of a stream flowing from a "spring" called mathematics. We believe that this sound will be quite different from the mathematics you learned before. We believe that you will gain more than a simple joy by learning mathematics in a fun and comprehensive way. We hope you find the freedom of spirit and flexibility of mind that lie deep within rational thinking. We also hope that you will start reading this book with an attitude of optimism and finish it with a greater understanding of both mathematics and the human mind.

And now for a synopsis of the contents of this book: in the first chapter of Volume I, we show that with the use of a limited number of mathematical equations, we can clarify the mysterious world of geometry hidden in curved surfaces. This clarification can be achieved by adopting a more global point of view in analyzing geometrical figures. This global point of view also will lead us towards the field called topology. This introduction to topology comes from the lectures of Shigeyuki Morita. At the end of Chapter 1, Lecture 2 is the proof of the Poincaré–Hopf theorem. The final step of this proof came from a student at one of Morita's lectures at the Tokyo Institute of Technology. This student approached Morita after the lecture and explained his proof method. Although Morita had shown a different method of proof in his lecture at the institute, he found the student's proof interesting and chose it for the book.

The second chapter addresses the topic of dimensions. You may find it interesting to know that we added this topic to the book as a result of many questions about the dimension asked at a gathering after one of our lectures. We have summarized these questions and presented them from a somewhat general point of view.

Chapter 3 in Volume II covers the journey down the road of the history of trigonometric functions. These functions were born from studying a circle. From a functional point of view they express the infinity hidden in circles. We finally reach the concept of power series expansion of functions. This road also leads to the birth of elliptic functions. This material comes from a lecture given by Koji Shiga.

The main theme of Chapter 4, which comes from the lecture of Kenji Ueno, is a geometric theorem called the Poncelet closure theorem. Among the many theorems of synthetic geometry developed during the nineteenth century, this theorem stands out for its extreme beauty of simplicity and depth. When you explore the relationship between ellipses and tangent lines which appear in the theorem from the point of view of algebraic geometry, you will arrive at the intersection point of geometry and algebra. These two distinctly different fields of mathematics influence, deepen, and expand upon one another. This relationship continues to influence modern mathematics. Let us walk together down this road to a heightened understanding of the mystical properties hidden in geometrical figures. We hope that at the end of our journey you will experience the joy of learning geometry.

You can read volumes and chapters of this book in any order; they do not have to be studied in sequence. We hope you will accept and enjoy our gift.

March 1995

Koji Shiga

Chapter 1

Invitation to Topology
(Viewing Figures Globally)

Introduction

For many people, topology is not a familiar term because it comes from a relatively new field of mathematics. Topology was developed in the twentieth century, while many other fields of mathematics, including calculus, had already been developed (at least in a primitive form) three hundred years ago. The word "topology" comes from the Greek—*topos* and *logos*—meaning location and study. When Poincaré began to explore this new field at the end of the nineteenth century, he did not call it topology, but instead called it Analysis Situs (analysis of situation). In fact, the origin of topology can most likely be traced to Euler's work in the eighteenth century, and that of Gauss and Riemann in the nineteenth century.

Topology is a discipline that is not restricted by traditional thinking. The way to approach it is, in essence, simply to observe geometrical figures carefully, and to grasp the totality of their image. There are no prerequisites to study this subject. Indeed, we can give lectures to show what topology is without using any complicated equations. When we say "observe geometrical figures carefully," we mean that we observe their important properties that do not change when the figure is continuously but slightly deformed, and so the resulting figure is considered to be the same as the original. In Figure 0.1, you can see figures of different shapes on the left. When we change the shapes of these figures slightly but continuously by stretching and contracting them, we eventually will obtain the figures on the right. In topology we consider these figures to be the same.

In geometry, which is the study of figures, the properties that have to be studied are determined by an agreement on which two figures should be considered to be the same. This is because geometry is the study of common properties of figures considered to be the same and of relationships between those properties. For example, if a figure is moved and can be placed on top of another figure in such a way that they "match," then these figures are viewed as the same, and they are called congruent. Lengths and angles are the characteristic properties of congruent figures. The geometry you may have learned thus far considers the properties of figures from the point of view of congruency.

NOTE. There are many different geometries. In Euclidean geometry, you can compare congruency by placing one figure atop the other. If they match, they are congruent. The length and the angles are the characteristics of geometric congruency.

FIGURE 0.1

Sometimes in geometry, similar (proportional) figures are compared. In such a case, one considers the angles and the ratios of lengths of figures. The type of geometry that studies figures from the point of view of congruency and similarity was established about 2400 years ago in Greece.

From the standpoint of the twentieth century mathematics, geometry can be roughly divided into two parts:

> (differential) geometry—where lengths and angles are considered important,

and

> topology—where slight differences in lengths and angles are considered insignificant.

However, these two fields of modern mathematics are not independent of one another. Rather, they represent two different ways of looking at figures. When our readers study a figure, the use of both methods allows them to see the properties of the figure more clearly. These two points of view enhance each other and often bring out hidden properties of figures that might not be visible when using one of these mathematical methods alone.

In this chapter, we will attempt to draw a picture of modern geometry by discussing the Euler characteristic of a surface.

LECTURE 1

The Euler Characteristic

1.1. The Euler Characteristic of a Square

We begin this lecture with partitioning a square into triangles. There is a rule about how a square is allowed to be partitioned into triangular pieces. The rule is that the pieces must fit together along the edges (see Figure 1.1).

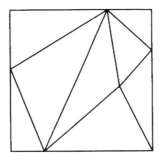

FIGURE 1.1

Given this rule, the following examples of improperly placed triangles would not qualify as a proper partition (see Figure 1.2). Under each figure there is an explanation of the "violation" of our rule.

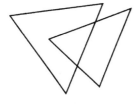

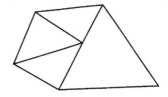

(a) No overlapping of triangles is allowed

(b) The vertex of a triangle cannot touch the edge of another triangle

FIGURE 1.2

Now count the number of each of the following elements in the partitioned square in Figure 1.1:

The number of triangular faces: f,

The number of edges: e,

The number of vertices: v.

5

You should find the following numbers:

$$f = 8, \quad e = 16, \quad v = 9.$$

Using these numbers, if we first subtract the number e of edges from the number f of faces and add the number v of vertices, we get

$$f - e + v = 1.$$

CHALLENGE 1. Partition a square into triangles in any manner you choose, following the rules in Figures 1.1 and 1.2, and count f, e, and v. Then find the number

$$f - e + v.$$

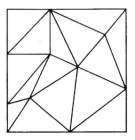

FIGURE 1.3

In general, a partition of a figure K on the plane or in space into triangles following the above-mentioned rules is called a triangulation of K. From a triangulation of the figure K, we find f, e, and v. The number $f - e + v$ is called the **Euler Characteristic** of K, denoted by $\chi(K)$, and we write

$$\chi(K) = f - e + v.$$

♣ Leonhard Euler (1707–1783) is considered by many to be the greatest mathematician of the eighteenth century. He is the author of many important works. He lost his eyesight shortly before his death; nonetheless, he continued to write articles.

There are many ways to triangulate a figure. If you change the pattern of partitioning, then the numbers f, e, and v also change. But no matter how you triangulate it, the number $f - e + v$ remains the same. We will not give a general proof of this fact in this book, but we will provide concrete examples to prove specific cases.

Consider a square. If you have completed Challenge 1, compare your answer with those of your neighbors. You should find that your answer is 1.

From our experiment, we might say that the number $f - e + v$ is the same for different triangulations. But what happens if you partition a square into one billion triangles? Then it will be almost impossible to count the numbers f, e, and v. Therefore, we need an exact proof that will guarantee that any triangulation of a square results in the Euler characteristic being 1. The next theorem provides the required proof.

THEOREM. $\chi(\text{square}) = 1$.

PROOF. Consider the triangulation of the square in Figure 1.1. Let us remove one of the triangles adjacent to the boundary of the square. The resulting figure is no longer a square. However, if we shrink and stretch the edges of the remaining triangles, we can deform the figure into a square again. Now we have a square with a different triangulation—with one triangle less—than the previous one.

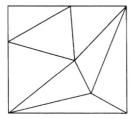 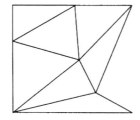 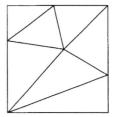

FIGURE 1.4

When we observe the process of deformation in Figure 1.4 carefully, we find that the triangulation of a square[1] is obtained by adding triangles one at a time as in the following two methods ((a) and (b) in Figure 1.5). In the above example, we removed one triangle from the existing triangulation, and in the following examples we add one triangle (shaded part).

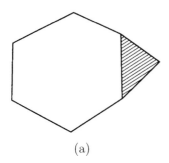 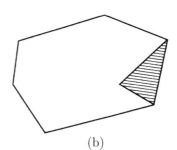

(a)   (b)

FIGURE 1.5

In the case of Figure 1.5(a), the numbers f, e, and v change, once a triangle is glued, as follows (the changes are indicated by arrows):

$$f \to f+1, \quad e \to e+2, \quad v \to v+1.$$

In the case of Figure 1.5(b), we have

$$f \to f+1, \quad e \to e+1, \quad v \to v.$$

---

[1]Remember that a square in topology is not necessarily the same as a Euclidean square.

However, in both cases, the number

$$f - e + v$$

is constant.

Any triangulation of a square is obtained as follows (see Figure 1.6). Starting from the triangulation obtained by partitioning the square into two triangles by a diagonal, we add some triangles (as in (a) and (b) above) and deform the resulting figure into a square. The original triangulation has $f = 2$, $e = 5$, $v = 4$, and hence $f - e + v = 1$. Therefore, for any triangulation we have

$$f - e + v = 1$$

This concludes the proof. $\square$

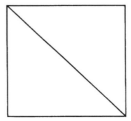

FIGURE 1.6

♣ Those of you who have studied mathematical induction, probably have noticed that we proved the above by induction. We used induction on f, e, and v starting with $f = 2$.

Once we know that the Euler characteristic of a square is 1, we also know that the Euler characteristic of any polygon on the plane is 1, that is,

$$\chi(\text{polygon}) = 1.$$

This is due to the following arguments. After we triangulate the polygon, we add new triangles to the triangulation and deform the resulting figure into a square without changing the number

$f(\text{number of triangles}) - e(\text{number of edges}) + v(\text{number of vertices}).$

In this way, we obtain the triangulation of a square (see Figure 1.7).

1.2. The Euler Characteristic of a Sphere and of a Torus

For the remainder of this discussion, let us imagine the triangles to be made of, say, rubber or Silly Putty®, with the ability to shrink and stretch freely. Now, we can make a sphere or a torus[2] by gluing several stretchy triangles edge to edge, in effect forming a patchwork sphere or torus out of these flexible triangles.

---

[2]In mathematics, the surface that has the shape of a doughnut, is called a torus.

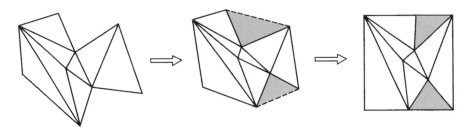

add shaded triangles

$$f \to f + 2$$
$$e \to e + 2$$
$$v \to v$$

FIGURE 1.7

If you find this hard to envision, think of a balloon or a soccer ball. This might make it easier to visualize how a sphere is partitioned into triangles (Figure 1.8).

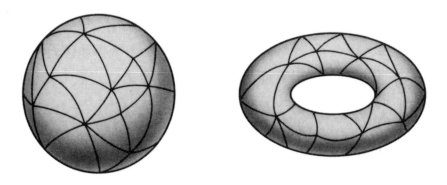

FIGURE 1.8

For such triangulations, what is the Euler characteristic of a sphere and a torus? The Euler characteristic of a polygon on the plane is 1, but the figures in Figure 1.8 are not on the plane. Does this affect their Euler characteristics? The answer is yes; the Euler characteristic does change, as indicated in the following theorem.

THEOREM. $\chi(\text{sphere}) = 2$,

$\chi(\text{torus}) = 0$.

Now, let us prove this theorem for a sphere. (It is somewhat more difficult to prove it for a torus; therefore, we will address this proof later on.)

Consider the triangulation of a sphere shown in Figure 1.9. We then remove one of the triangles. The resulting figure is a sphere with a missing piece shaped like a triangle. Since this figure is composed of stretchy triangles, we can now stretch it into a flat figure on the plane.

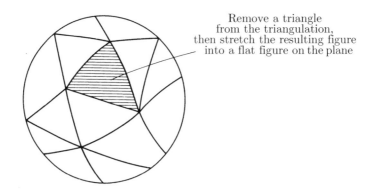

Remove a triangle
from the triangulation,
then stretch the resulting figure
into a flat figure on the plane

FIGURE 1.9

♣ If you find this difficult to imagine, try the following trick. First, stretch and shrink the triangles on a completely intact sphere without any triangles removed. Then stretch one particular triangle (to be removed later on) to the size of approximately one half of the sphere, and shrink the remaining triangles accordingly. Finally, remove the "stretched" triangle. The resulting figure will look like a bowl. Now it is easy to see how to flatten the bowl in such a way that it becomes a flat figure on the plane.

Let us call this flattened figure K. It is now a triangle whose perimeter matches the three sides of the removed triangle, and K is already triangulated (Figure 1.10).

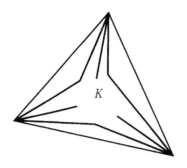

FIGURE 1.10

The Euler characteristic of the figure K is 1 because it is a polygon.

Comparing the triangulation of K and that of the original sphere, we see that the numbers e (edges) and v (vertices) have not changed. However, the number of faces has increased by one. Therefore,

$$\chi(\text{sphere}) = \chi(K) + 1$$
$$= 1 + 1 = 2.$$

From this proof, we can see that the Euler characteristic of a sphere does not depend on triangulation (since $\chi(K)$ does not depend on triangulation).

We have shown that the Euler characteristic of a sphere is 2. Now let us turn to a torus.

As we stated earlier, it is more difficult to show that the Euler characteristic of a torus does not depend on triangulation.

If we remove a triangle from a torus and attempt to spread the resulting figure flat on the plane, we will find this impossible to do (unlike our experience with a sphere). Hence, we must approach this problem from a slightly different angle. First, consider an elastic rectangle $ABCD$. We make a tube by gluing the opposite sides AB and DC of the rectangle (Figure 1.11). Then we glue the circular ends of the tube to form a torus (Figure 1.11). Thus, we obtain a torus by considering the corresponding points on the sides AB and DC to be the same and AD and BC to be the same, that is, by identifying the opposite sides of the rectangle.

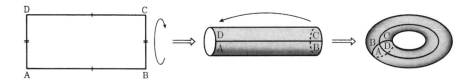

FIGURE 1.11

Now let us triangulate the rectangle as in Figure 1.12. As you already know, the Euler characteristic of a rectangle is 1. However, if we identify the pair of opposite sides of the triangle, the number of edges is four less than the number of edges counted before the identification, and the number of vertices is five less. Hence, the Euler characteristic of a torus is (using this particular triangulation)

$$\chi(\text{rectangle}) - (-4) + (-5) = 1 + 4 - 5 = 0.$$

In the case of a torus, we considered a particular triangulation, so, as we mentioned before, our proof is not yet complete. However, we have at least established that the Euler characteristic of a sphere is 2, and that of a torus is 0. In other words, they are not the same. To prove that the Euler characteristic of a torus is 0 for any triangulation, we must consider a more general case. It is better to prove our assertion in the context of surfaces with many holes. When we consider this kind of surface, it is useful to

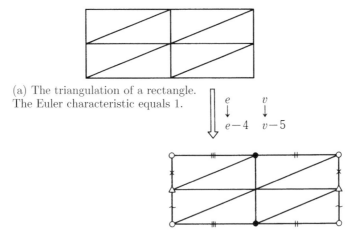

(a) The triangulation of a rectangle.
The Euler characteristic equals 1.

$$e \quad\quad v$$
$$\downarrow \quad\quad \downarrow$$
$$e-4 \quad v-5$$

(b) Edges and vertices with identical
symbols are considered the same.
After this identification the
Euler characteristic is zero.

FIGURE 1.12

identify edges of a flat figure just as we did previously to obtain a torus.
Now, let us talk about a surface with many holes.

1.3. Closed Surfaces

A bounded smooth surface without boundary is called a closed surface.

♣ In Figures 1.13 and 1.14, we show examples of surfaces
which are not closed. The surface in Figure 1.13, which ex-
pands infinitely, is not bounded. Therefore, it is not a closed
surface.

FIGURE 1.13

The surface in Figure 1.14 (a) has an edge (boundary points).
Therefore, this surface is not a closed surface.
The surface in Figure 1.14 (b) is jagged, not smooth. There-
fore, this surface also is not a closed surface.

In topology, we consider two figures to be different if one cannot be
transformed into the other by stretching and shrinking. From this topological point of view, closed surfaces are completely classified. To put it simply,
they are given by "g-man life savers." Imagine a life saver built for two
people, that is, with two holes. This two-man life saver would be a closed

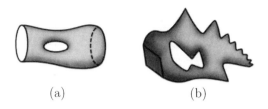

(a)   (b)

FIGURE 1.14

surface. The same is true for a three-man life saver or any "g-man life saver" $(g = 1, 2, 3, 4, \ldots)$ (see Figure 1.15).

The "g-man life saver" is called a closed surface of genus g and denoted by S_g:

S_0 is a life saver without a hole (a sphere).
S_1 is a regular life saver with one hole (a torus).
S_g is a life saver with g holes.

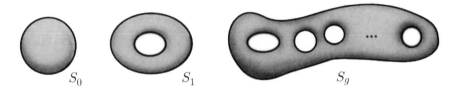

FIGURE 1.15

Using these notation, let us write the Euler characteristics of the closed surfaces we have studied so far. They are

$$\chi(S_0) = 2$$

and

$$\chi(S_1) = 0.$$

Given the above, we might ask ourselves, "What is the number $\chi(S_g)$?"

We will consider this problem in the case of a closed surface S_2 of genus 2. In the same way as we obtain a torus by identifying the opposite sides of a rectangle, we should be able to obtain S_2 by identifying appropriate sides of an octagon. We will explain this idea in Figure 1.16 by drawing the surfaces on the left and their polygonal representations on the right.

First, let us observe what happens if we puncture a hole in the surface of the torus.

Using the above polygonal representation, we will obtain the polygonal representation of S_2 by cutting it up and then gluing it back together as in Figure 1.17.

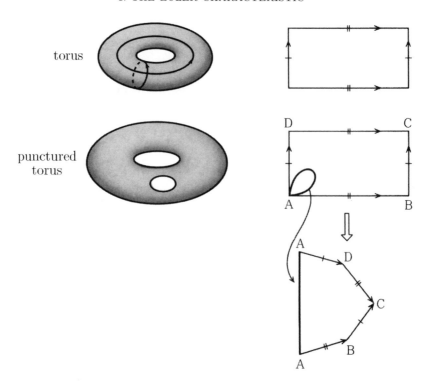

torus

punctured
torus

FIGURE 1.16

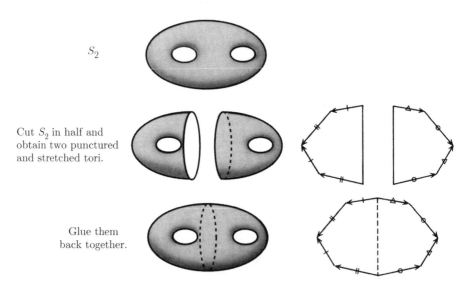

S_2

Cut S_2 in half and
obtain two punctured
and stretched tori.

Glue them
back together.

FIGURE 1.17

By labeling the sides to be identified with letters α and β with subscripts,
we can obtain the polygonal representation of S_2 as an octagon shown in
Figure 1.18.

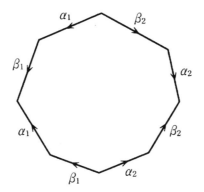

FIGURE 1.18

The edges of an octagon, α_1, β_1, α_2, β_2, appear as curves on the surface S_2 shown in Figure 1.19.

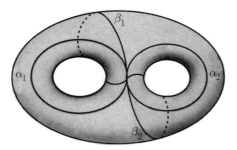

FIGURE 1.19

Once a polygonal representation of S_2 is obtained, we can calculate the Euler characteristic of S_2 using the triangulation of the octagon shown in Figure 1.20. In this case,

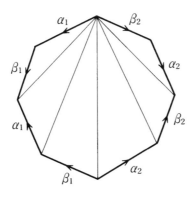

FIGURE 1.20

$$f = 6, \quad e = 9, \quad v = 1.$$

Therefore, the Euler characteristic for this triangulation is

$$\chi(S_2) = 6 - 9 + 1 = -2.$$

CHALLENGE 2. Verify that the Euler characteristic of S_2 is -2 (that is, $\chi(S_2) = -2$) using the polygonal representation.

We have the numbers $\chi(S_1) = 0$ for a torus, and $\chi(S_2) = -2$ for a closed surface of genus 2 obtained from looking at particular triangulations. If we consider the case of S_2, could it be possible to get another number, say, -3 for a different triangulation? The answer is no. So far, we have not established that the Euler characteristic is a characteristic property of a surface. However, for any triangulation, the Euler characteristic of a closed surface is constant. It is independent of the choice of triangulation.

We will use the example of the surface S_2 to explain how to prove this important fact using the concept of barycentric subdivision. The barycenter of a triangle is the point where all three medians of the triangle meet, that is, the intersection point of the lines drawn from each vertex to the midpoint of the opposite side of the triangle (see Figure 1.21).

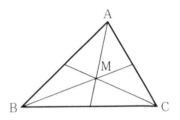

FIGURE 1.21

For any triangulation of S_2 we connect the barycenter of each triangle with the vertices and midpoints of the sides of this triangle. Then each triangle will be divided into six smaller triangles. Thus, we obtain a finer triangulation of S_2 (see Figure 1.22). This operation is called a **barycentric subdivision**.

The Euler characteristics of S_2 does not change after a barycentric subdivision. This fact can be demonstrated using the following table:

Before Barycentric Subdivision	After Barycentric Subdivision
$f \Rightarrow$	$6f$
$e \Rightarrow$	$2e + 6f$
$v \Rightarrow$	$v + e + f$
$f - e + v \Rightarrow$	$6f - (2e + 6f) + (v + e + f)$
	$= f - e + v$

CHALLENGE 3. Verify the above results.

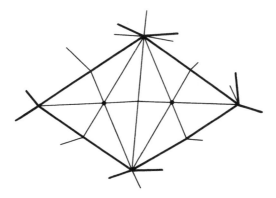

FIGURE 1.22

Now suppose a triangulation of S_2 is given. We denote it by Δ. Since we can perform a barycentric subdivision repeatedly, we can obtain finer and finer triangulations as shown in the diagram below.

The triangulation of S_2: $\Delta \to \Delta^{(1)} \to \Delta^{(2)} \to \cdots \to \Delta^{(k)} \to \cdots$

($\to$ indicates barycentric subdivision).

To find the Euler characteristic of S_2 for the triangulation Δ, it is sufficient to show that the Euler characteristic of S_2 for $\Delta^{(k)}$ is -2, since the Euler characteristic does not change after a barycentric subdivision.

This means that we can assume that each triangle of the given triangulation is sufficiently small. As we shall see, by using this triangulation, we obtain the Euler characteristic of S_2 to be -2.

First, draw a polygonal representation of S_2, as in Figure 1.18. Then draw four curves α_1, α_2, β_1, β_2 on the surface S_2, as in Figure 1.19. Since each triangle that appears in this triangulation is sufficiently small, we can assume that if a side of a triangle intersects a curve, then this side does not intersect another curve. In other words, each side of a triangle in the triangulation intersects at most one curve. This is possible since we previously ensured that each triangle is sufficiently small. Now each side of the octagon can be approximated by the jagged line made of the sides of the triangles which intersect this side of the octagon. The totality of these jagged lines forms an n-gon that follows the outline of the octagon and passes through each vertex of the octagon. On any two sides of the octagon to be identified there will be jagged lines which should also be identified (Figure 1.23).

Consider this n-gon. We can see that the triangulation of S_2 gives rise to a triangulation of this n-gon. Now, identify the corresponding sides. Let f_0, e_0, and v_0 be the numbers of faces, edges, and vertices before the identification. Then, we have

$$f_0 - e_0 + v_0 = \chi(n\text{-gon}) = 1.$$

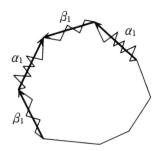

FIGURE 1.23

Next, let f, e, and v be the numbers of faces, edges, and vertices after the identification (these are the actual f, e, and v we need to calculate for S_2). Then

$$f = f_0,$$

$$e = e_0 - \frac{1}{2} n,$$

$$v = v_0 - 7 - \frac{1}{2}(n - 8).$$

Therefore, the Euler characteristic of S_2 is given by

$$f - c + v = f_0 - e_0 + v_0 + \frac{1}{2} n - 7 - \frac{1}{2} n + 4 = 1 - 3 = -2.$$

This means that for any triangulation of S_2 the Euler characteristic of S_2 is -2. Hence, the equality $\chi(S_2) = -2$ is established. Similarly, for any triangulation of a torus the Euler characteristic is 0. Hence $\chi(S_1) = 0$.

1.4. Closed Surfaces and Their Euler Characteristics

So far, we have learned that for the closed surfaces S_g $(g = 0, 1, 2, \dots)$, $\chi(S_0) = 2$, $\chi(S_1) = 0$, and $\chi(S_2) = -2$. As you can see, the Euler characteristics of S_0, S_1, and S_2 are decreasing by 2. You might have guessed that $\chi(S_3) = -4$, $\chi(S_4) = -6$, and $\chi(S_5) = -8$. If you did, you would be right because of the following theorem.

THEOREM. $\chi(S_g) = 2 - 2g$.

Let us first prove this theorem. Recall the discussion of S_g as a life saver for g people. Since we observe the surface from a topological point of view, we can change the surface continuously and obtain a new look of the resulting surface. In Figure 1.24, we show an example of a continuous deformation of S_3.

For the surface S_g obtained by this kind of deformation, we can use the polygonal representation shown in Figure 1.25. Take a few minutes to examine this fact; it is an important one.

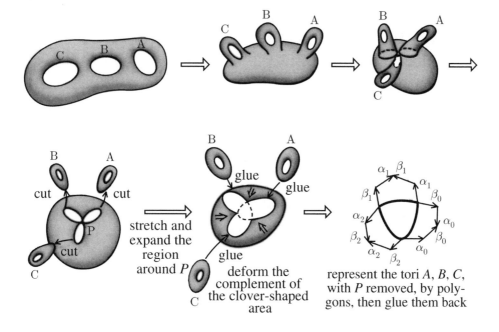

FIGURE 1.24

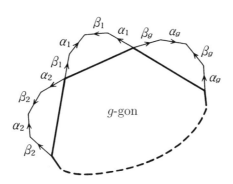

FIGURE 1.25

From this we can obtain the standard polygonal representation of S_g as in Figure 1.26. This is a generalization of a polygonal representation of S_2. The standard polygonal representation of S_2 is an octagon. The standard polygonal representation of S_g is a $4g$-gon. In this $4g$-gon, sides are glued together in pairs with opposite orientations. Figure 1.26 shows the lines which give a triangulation of S_g.

Let us now find the Euler characteristic of S_g for this particular triangulation. We have

$$f = 4g - 2, \quad e = 4g - 3 + 2g, \quad v = 1,$$
$$f - e + v = (4g - 2) - (4g - 3 + 2g) + 1 = 2 - 2g.$$

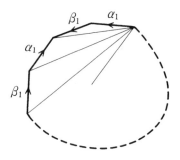

FIGURE 1.26

Hence, we have proved the theorem for a particular triangulation.

We can prove that the Euler characteristic does not depend on the choice of triangulations by using the same method we used for S_2.

Now we know that we can completely classify the closed surfaces by their Euler characteristics even if they appear drastically different after continuous deformation. The classification is shown in Figure 1.27.

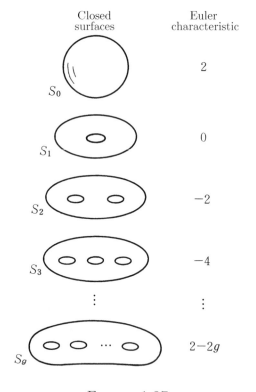

FIGURE 1.27

By this time, you likely have some fundamental questions, not least of which is whether the Euler characteristic of a surface is a different way of counting the number of holes in the surface. Another question might be: if we probe even deeper, will we be able to find a surface with Euler characteristic 1? We cannot give a satisfactory answer to the first question in this book. Actually, the concept of "holes" in a life saver made for many people is quite vague from the topological point of view. So far, we have been talking about surfaces which look like a life saver for g people. However, a surface such as the one shown in Figure 1.28 (which looks as if it was created by an avant-garde sculptor) is in fact a closed surface.

QUESTION. How many holes are there in the surface in Figure 1.28 and how shall we count them?

FIGURE 1.28

These questions may puzzle you. But if you use the Euler characteristic, all you need to do is to count the number of faces, edges, and vertices of triangles "pasted" on this sculpture. So we can obtain the Euler characteristic of this surface by using triangulation. The Euler characteristic of this surface is in fact -2. Therefore, from the topological point of view this surface is considered the same as S_2. However, you cannot obtain this surface from a soccer ball with two holes by continuously deforming it unless self-intersections are allowed. Actually, if we deform a soccer ball in four-dimensional space instead of the three-dimensional space in which we live, we can obtain this surface without self-intersection. We will discuss this again in Chapter 2, "The Story of Dimension". In any case, the Euler characteristic provides a method to classify closed surfaces by studying an inherent property of these surfaces instead of just visual observations. The Euler characteristic gives us a tool for enhancing the study of these mathematical objects.

Next, let us turn to the second question. If we consider a surface P called projective plane, we know that $\chi(P) = 1$. However, we cannot produce a projective plane as a non-self-intersecting surface in three-dimensional space. We will further discuss projective planes in Chapter 2, but for now, assume that a projective plane is a surface obtained by identifying the opposite sides AB and CD, and AD and CB of a rectangle $ABCD$ shown in Figure 1.29.

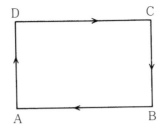

FIGURE 1.29

CHALLENGE 4. Consider the triangulation of a projective plane given in Figure 1.30. Find the numbers f, e, and v and show that

$$\chi(P) = 1.$$

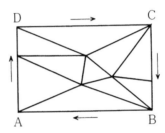

FIGURE 1.30

We have constructed the torus and projective plane by identifying the opposite sides of a rectangle. We can construct yet another surface by yet another identification. This identification, shown in Figure 1.31, gives the famous Klein bottle, which makes no distinction between inner and outer surfaces. Perhaps you have heard about the Klein bottle in a science fiction story.

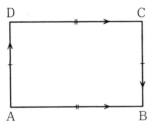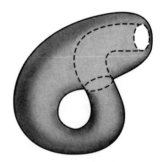

FIGURE 1.31

CHALLENGE 5. What is χ(Klein bottle)?

♣ Suppose an ant falls into a Klein bottle and begins crawling on the "inner" wall of the bottle. Eventually he will find himself crawling on the "outer" wall of the bottle. The Klein bottle will not be a good ant's nest. To show how puzzled the ant would be, we present a more detailed picture of a Klein bottle in Figure 1.32.

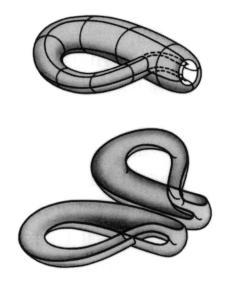

FIGURE 1.32. Klein bottle cut in half.

Vortices Created by Winds
and the Euler Characteristic

2.1. Directions of Winds and Centers of Vortices
(Vector Fields and Critical Points)

At any given time, winds are blowing in many places of the earth. At the same time, there also should be places on the earth where there is no wind. To further explain this idea, we can say that if there were no place on the earth where there is no wind, then there would definitely exist places where the flow of the wind becomes discontinuous, like a tornado. For example, try to imagine the situation when the west wind blows at every point on the earth. Under this imaginary conditions, there would actually be two places where there is no wind, namely the North Pole and the South Pole. If, however, you were a resident of a planet shaped like a doughnut, it could happen that wind blows in the same direction at any point on the planet with a constant velocity (Figure 2.1). This sounds like science fiction, doesn't it?

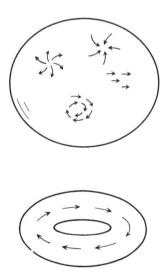

FIGURE 2.1

The source of these differences can be explained by the following argument. The Euler characteristic of the surface of the earth, which looks like a sphere, is 2. The Euler characteristic of the surface of a doughnut, which is a torus, is 0. In this lecture, our main theme is to determine how many centers of vortices exist in whirlwinds created by winds blowing on such surfaces, and to study how that number is related to the Euler characteristic of these surfaces.

We started our lecture with discussing winds, but we could have started out with looking at flows of water on surfaces. In such a situation, we express a flow of water by vectors given at each point on the surface and indicating the direction of the flow. Mathematically, we start by assuming that at each point on the surface, a vector showing the direction of the flow is given. Such a vector is called a **tangent vector**.

We assume that vectors given at points on a surface change continuously. If we discus this in the "language" of flows, we are talking about a continuous flow. If we are talking in the "language" of flows of wind, we are assuming that there is no sudden change of wind, such as a tornado, which blows vertically. If we are talking about flows of water, we are assuming there is no sudden movement of water such as ocean currents colliding with each other and creating a discontinuity of the flow.

As we mentioned before, if a tangent vector is given at each point on the surface and if the direction and magnitude of the vector change continuously from point to point, we say that a vector field (the set of tangent vectors) is given on the surface. To understand this, imagine that our surface is a field with alternating regions of concavity and convexity and with small vectors lying on it, each differing slightly in direction. This explanation may help to clarify the words "vector field".

Now, at the point where the flow stops (in the case of wind, this is a point of no wind), the corresponding tangent vector is a zero vector. Such a point is called a **critical point** of the vector field.

NOTE. So far, we were talking about definitions. If you are having difficulty understanding what we have been discussing, we will show you concrete examples. Don't be concerned; keep reading!

Let us explain the relation between flows and vector fields using the example of rain falling on the earth (Figure 2.2). We indicate the directions of the flow of rainwater by vectors (precisely speaking, the vectors are velocity vectors expressed as arrows). Let us examine how the critical points appear in this case.

Using the flow diagrams in Figure 2.3, which show how the flows of water create critical points at the peak of a hill, at the low point of a valley, and at a pass. We can tell that at the peak and at the lowest point of the valley, the directions of the vectors at the critical points are opposite with respect to the critical points. The flow of water on the pass is divided by the pass itself towards different valleys.

Rainfall

(a)

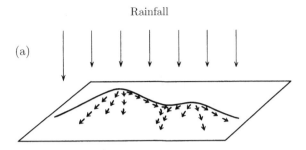

The flow of rainwater is represented by vectors ($\downarrow\uparrow$)

(b)

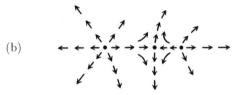

The dot (•) indicates a critical point

FIGURE 2.2

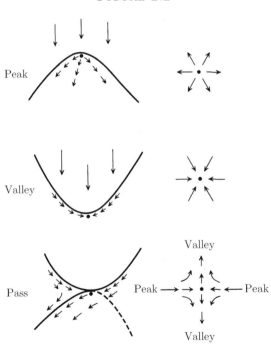

FIGURE 2.3

To determine which types of critical points exist in the flow on a surface
we focus only on the flow near critical points and examine the flow of water

that enters the critical points and the flow of water that comes out from the critical points. We do not need to draw the entire picture of a surface since we examine only the situation near critical points. Different types of critical points are shown in Figure 2.4. If we examine the flow on the surface, we will find one or more of these critical points.

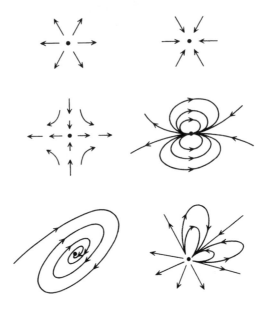

FIGURE 2.4

2.2. The Index of a Critical Point

If the top of a mountain is as flat as the top of a table, then rain falling on the top of the mountain will collect on this flat area rather than running off as it would from a normally sloped mountain top. In such a case, the critical points are everywhere on the "flat" top of the mountain. From now on, we will exclude such a case. We will consider a critical point to be a point around which there always exists a flow of water in certain directions (in mathematics, we call such a critical point an isolated critical point).

Now, we turn our attention to the flow of water entering or leaving a critical point, and with each critical point we associate an integer called the **index** of the critical point. The idea is as follows. Draw a circle centered at the critical point and such that no other critical point lies inside the circle or on the circle. To make the explanation simpler, consider the critical point as the center of the vortex of a storm. Assume that a man starts at an arbitrary point on the circle and walks counterclockwise along the circle until he returns to the starting point. (The wind, however, is not necessarily blowing counterclockwise; it could be coming from all directions.) If the direction of the wind at any given point is from the east, west, north, or south, then while walking along the circle, this man may feel an east wind in his face at some moment and a south wind in his face at some other moment. At other times, he may feel a sudden change in the direction of the wind after taking only a few steps. If we represent the directions of the wind the man feels while walking, by arrows of unit length whose tails are positioned at the origin, then the tips of the arrows lie in the unit circle[1] centered at the origin. While the man is walking along the circle, the tips of the arrows move back and forth on the unit circle. If the flow of the wind is very erratic, then the tips of the arrows oscillate erratically. However, when the man returns to the starting point, the arrow will be pointing in the same direction as it was when he first started walking.

If it is still not clear, imagine a man carrying a weather vane with him as he walks along the circle (starting at an arbitrary point). The weather vane will move to indicate the direction from which the wind is blowing. When the man returns to the point from which he started, the weather vane will be pointing in the same direction as it was when the man started walking.

Two simple examples are shown in Figure 2.5. In part (I), the wind

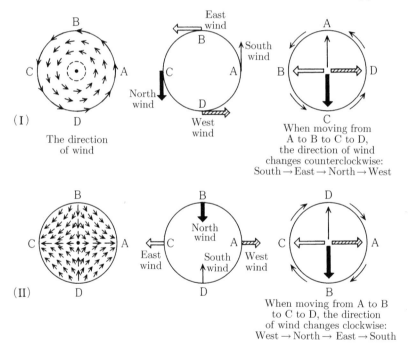

FIGURE 2.5

blows counterclockwise around a critical point P. Therefore, the directions

[1]On the coordinate plane, a circle with center at the origin and radius 1 is called the unit circle. The standard equation of the unit circle centered at the origin is given by

$$x^2 + y^2 = 1.$$

of the wind represented by the arrows as described above revolve once counterclockwise, and we say that the index of the critical point P is 1. In part (II), the wind flow is more complicated. In that case, the arrows make one complete turn in the clockwise (negative) direction. In this case, we say that the index of the critical point P is -1.

Hopefully, our readers have grasped the general concept of the index of the critical point. Now, let us substitute the word "vector" for the phrase "flow of the wind" and define the index of the critical point P, denoted $i(P)$, as follows. Draw a circle centered at the critical point P and while going along this circle counterclockwise, observe the change of the direction of the vector. If the direction of the vector revolves counterclockwise n times, then we define $i(P) = n$. If the direction of the vector revolves clockwise n times, we define $i(P) = -n$.

Having said this, does it mean that there are cases in which the index of P is not only 1 or -1, but 2 or 3? Or even -2 or -3? The answer is yes. We will show examples of such critical points and their indices in Figure 2.6.

CHALLENGE 6. For indices 3 and 4, examine how the direction of the vectors changes.

CHALLENGE 7. For indices -3 and -4, examine how the direction of the vectors changes.

From the examples in Figure 2.6, you might have noticed that if an "elliptic flow", shown in Figure 2.7, is added to the existing flow, then the index of the critical point increases.

On the other hand, if a "hyperbolic flow" is added, as in Figure 2.8, then the index of the critical point decreases. Using these observations, we can create a flow with a critical point of index n for any integer n.

2.3. The Poincaré–Hopf Theorem

More than one hundred years ago, at the end of the 19th century, Henri Poincaré proved the following theorem.

POINCARÉ–HOPF THEOREM. *Let S_g be a closed surface. For any vector field on S_g with finitely many critical points, the sum of the indices of the critical points is equal to the Euler characteristic of S_g (i.e., $\chi(S_g) = 2 - 2g$).*

In 1926, Hopf generalized this theorem and proved that it applies not only to closed surfaces, but also to general higher-dimensional smooth surfaces. Actually, this generalized theorem is called the Poincaré–Hopf theorem.[2] The theorem stated above is a simplified version of it. For convenience, by the Poincaré–Hopf theorem we mean this simplified version.

---

[2]The actual Poincaré–Hopf theorem states that the Euler number of a compact manifold M is equal to the sum of the indices of the critical points of any vector field on M which has only isolated critical points. (See Y. Choquet-Bruhat, C. DeWitt-Morette, and M. Dillard-Bleick, *Analysis, Manifolds, and Physics*, North-Holland, Amsterdam, 1982.)

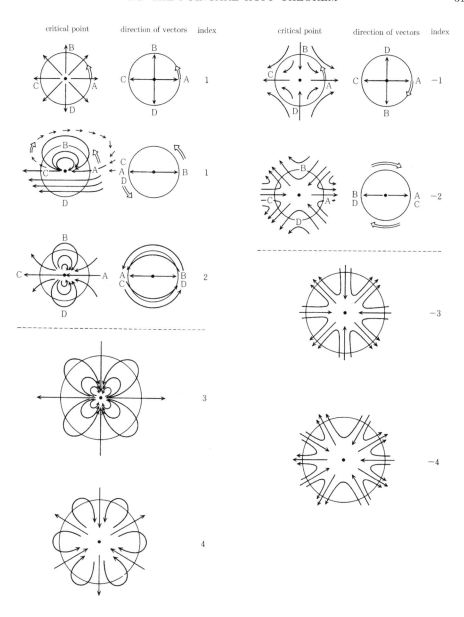

FIGURE 2.6

FIGURE 2.7

FIGURE 2.8

According to our discussion, the vector field in the above theorem can be referred to as a "flow". Therefore, we now explain by some examples what is the assertion of this theorem for "flows". We will prove the theorem in the next section.

First, if the surface S_0 is a sphere, then

$$\chi(S_0) = 2.$$

What does the Poincaré–Hopf theorem tell us about this surface? It shows us the following intuitive fact. If a wind is blowing over a sphere (without any significant discontinuity, such as a tornado), then there is always some place on the sphere where there is no wind (Figure 2.9).

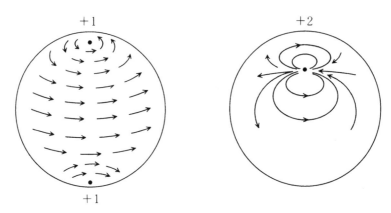

The flow of west winds

FIGURE 2.9

According to the Poincaré–Hopf theorem, we know that the sum of the indices of the critical points of a flow over the sphere is 2. This means that there must be at least one critical point (where there is no wind) on the earth.

For a slightly different example of a vector field on a sphere, consider the vector field shown in Figure 2.10, in which rainwater is falling from the top of a mountain to its base. Imagine the "mountain" in Figure 2.10 to be made of a pliable material. Now gather up the base of the mountain to one point, forming a ball. Thus, we obtain a vector field on a sphere. This vector field has four critical points: two peaks A and B, where the index of each point is $+1$, a pass C, where the index is -1, and the point D, at which the base was gathered, where the index is $+1$. Hence, the sum of the indices is $1 + 1 - 1 + 1 = 2$.

Next, if the surface S_1 is a torus, then

$$\chi(S_1) = 0.$$

In this case there can exist a flow of wind over this surface which has no critical points (unlike the wind flow on the sphere, where there is always at

(a)

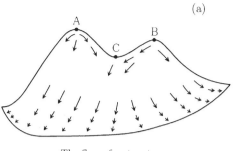

The flow of water stops
at the base of the mountain

(b)

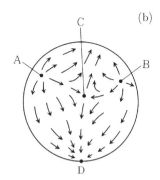

Point D where the base of the
mountain has been gathered up

FIGURE 2.10

least one point with no wind). On the torus the wind can blow everywhere. Such vector fields are shown in Figure 2.11. These examples demonstrate that there are two possible wind flows which have different directions on the torus.

FIGURE 2.11

It might be interesting to verify that the Poincaré–Hopf theorem holds on a torus for the vector field that is created by a rainfall (see Figure 2.12). This vector field has four critical points, and the sum of their indices is 0, as the theorem states.

The same is true for the general surface S_g (see Figure 2.13).

We can also construct a vector field from a triangulation of S_g. To obtain this vector field, first triangulate S_g, then perform barycentfic subdivisions over the triangulation. Next, for each original triangle, create the flow from each vertex to the midpoints of the sides, and to the center of the triangle, and from the midpoints of the sides to the center of the triangle as shown in Figure 2.14. Then extend the flow continuously to the faces of the triangles. In this way we construct a vector field on S_g. The critical points of this vector field appear only at the vertices, centers of the triangles, and midpoints of the sides of the triangles. The indices of the critical points are as shown in Figure 2.14.

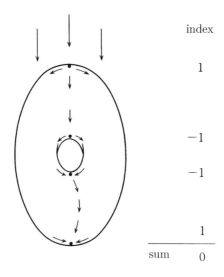

FIGURE 2.12. Vector field created by a rainfall on the torus.

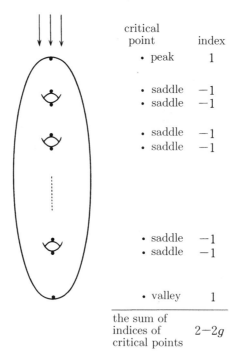

FIGURE 2.13. Vector field created by a rainfall on the surface of genus g.

Therefore, the sum of the indices of the critical points is

$$f - e + v = \chi(S_g).$$

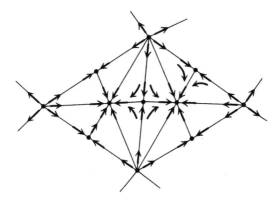

FIGURE 2.14. The index at the center of a triangle is 1; the
index at the midpoint of an edge is −1; the index at a vertex
is 1.

The Poincaré–Hopf theorem holds true again. It is interesting to see how
the Euler characteristic emerges naturally in this example.

2.4. Proof of the Poincaré–Hopf Theorem

Now we will show how to prove the Poincareè–Hopf theorem. Let us
briefly discuss the idea behind the proof. To do that, we need to discuss in
a little more detail the notion of winding vectors along a circle, about which
we talked when we defined the index of a critical point of a vector field.

Let $P(t)$ $(0 \leq t \leq 1)$ be a variable point on a circle C. When $t = 0$, $P(t)$
is at a starting point A on C; then $P(t)$ travels along the circle C, and at
$t = 1$, $P(t)$ is at the point A again. With this in mind, let us define intuitively
the "winding number" of the variable point $P(t)$ as follows. Imagine taking
a string of an appropriate length and trying to wind it around a circle C.
To do this, first fix one end of the string at point A, then align the loose
part of the string along the circle following the motion of $P(t)$. When the
string reaches A again at $t = 1$, cut off the remainder of string that reaches
past A.

For example, if we wind the string tightly twice in the positive direction
(counterclockwise), then it should look like the example in Figure 2.15. We
define the winding number of $P(t)$ as 2, which is the number of counterclockwise revolutions. If the string is wrapped around C clockwise, then we
define the winding number to be −2.

However, if the string is wrapped around C as in Figure 2.16, how should
we define the winding number of $P(t)$? In order to define the winding
number, we first pull the end of the string at $P(1)$ and take up all the slack.
We will find that the string is wound two times in the positive direction. In
this case we define the winding number to be 2. In essence, we take up the
slack from the string, wind it tightly, and count the winding number of the
tightened string. If it is wound in the positive direction, the winding number

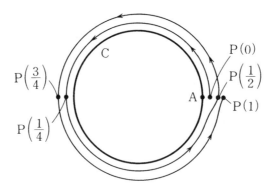

FIGURE 2.15

is a positive integer. If it is wound in the negative direction, the winding number is a negative integer. If we pull the string tight and it comes off the circle (i.e., it cannot completely go around the circle at least once), then it is not wound and we define the winding number to be 0.

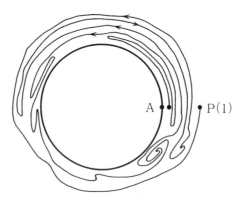

FIGURE 2.16

CHALLENGE 8. Find the winding numbers for $P(t)$ as in Figure 2.17.

From the above definition, the following two statements (a) and (b) hold.

(a) Winding numbers are integers, and a small change in winding does not cause a change in the winding number.

For example, if the string is wound two times around the circle, a small change of winding by allowing some slack or creating extra tension does not change the winding number (see Figure 2.18).

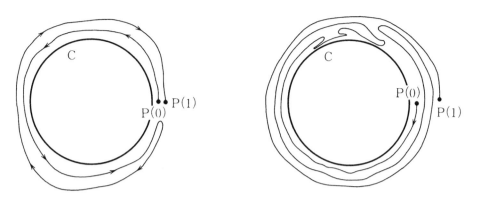

FIGURE 2.17

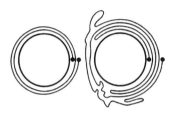

FIGURE 2.18

(b) Suppose we have two strings $P(t)$ and $Q(t)$ with winding numbers m and n, respectively. If we connect the endpoint $P(1)$ of $P(t)$ with the starting point $Q(0)$ of $Q(t)$ we obtain a winding of the combined string as follows:

$$P(0) \to P(1) = Q(0) \to Q(1).$$

Then the winding number of the combined string is $m + n$.

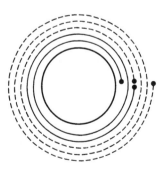

FIGURE 2.19. When two strings with winding numbers 2 and 3 are connected, the winding number of the resulting string is 5.

Using the above definition, we can say that the index of the critical point P of a vector field denoted by $i(P)$ is the winding number of the variable point $P(t)$ on a unit circle. This variable point $P(t)$ is obtained by first drawing a circle around the critical point P and then checking the change of the direction of the vector field on the circle. The change in the direction of the vector field at each point on the circle is represented by a point $P(t)$ on the unit circle.

    ♣   The direction of the vector field at each point on the circle can be expressed as a vector. If we take the initial point of the vector and place it at the center of the unit circle, then the endpoint of the vector will lie on the unit circle. We consider this endpoint to be the variable point $P(t)$.

Now let us consider the case where a vector field on a surface has three critical points P, Q, and R. We assume that P, Q, and R are located sufficiently close to one another. If we draw circles centered at P, Q, and R, then the sum of the winding numbers of the vector field is the same as the sum of the indices of P, Q, and R, that is, $i(P) + i(Q) + i(R)$. However, because the winding number has a property (a), a small deformation of the circle (i.e., slight change of the direction of the vector field on the circle) does not change the winding number. Keeping in mind this fundamental property of the winding number, let us study the illustrations in Figure 2.20 following the direction of the arrow.

At the symbol ($\star$), three deformed circles meet at a point, and the winding number of the vector field on the outer loop is $i(P) + i(Q) + i(R)$, as follows from property (b). Furthermore, if the loop is deformed into a circle, the winding number of the vector field on this circle coincides with the sum of the indices of the critical points.

With an understanding of the above facts, we can give the proof of the Poincaré–Hopf theorem. Consider a vector field on a surface S_g with critical points P_1, P_2, ..., P_s. For reference we will call this vector field (I).

We can assume that these critical points are all in one circle C, since from a topological point of view, we can take S_g to be made of a thin rubber film. The disk that contains the critical points P_1, P_2, ..., P_s could be made very small, while the rest of the surface could be enlarged. When we shrink and enlarge the surface as described above, the indices of the critical points do not change. This is due to property (a) of winding numbers.

We can assume that all critical points are located inside the circle C. Then, according to the above-mentioned facts, we can conclude that the sum of the indices of the critical points of the vector field (I) is given as follows.

(Ĩ) $i(P_1) + i(P_2) + \cdots + i(P_s) = $ the winding number of the vector field on the circle C.

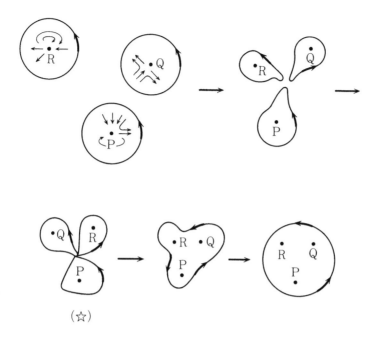

FIGURE 2.20

We already know that there exists a vector field on a surface S_g such that the sum of the indices of all its critical points is $2 - 2g$. For example, both the vector field constructed by the flow of rain and the vector field obtained by the barycentric subdivision of the triangulation of the surface have this property. (See Section 2.3.)

To prove the Poincaré–Hopf theorem, either one of the above vector fields will suffice, but let us use the vector field constructed by the flow of rainwater. We will refer to this vector field as (II). This vector field has $2 + 2g$ critical points (two of index 1 and $2g$ of index -1). Let $Q_1, Q_2, \ldots,$ Q_{2+2g} be these critical points. Stretch and shrink S_g and draw a circle C' containing all the critical points inside. Now deform S_g again so that C' coincide with the circle C mentioned above and, this time, let $Q_1, Q_2, \ldots,$ Q_{2+2g} also lie inside C. Similarly to relation $(\tilde{\mathrm{I}})$, we can conclude that

---

$(\tilde{\mathrm{II}})$ $i(Q_1) + i(Q_2) + \cdots + i(Q_{2+2g}) = 2 - 2g$

$\qquad\qquad$ = the winding number of the vector field constructed
$\qquad\qquad$ on the circle C from the flow of rainwater.

Therefore, if we can use relations $(\tilde{\mathrm{I}})$ and $(\tilde{\mathrm{II}})$ to show that

---

$(\#)$ the right-hand side of relation $(\tilde{\mathrm{I}})$

$\qquad\qquad$ = the right-hand side of relation $(\widetilde{\mathrm{II}})$,

---

then we will prove that the left-hand side of relation $(\tilde{\mathrm{I}})$ is $2 - 2g = \chi(S_g)$. This is the Poincaré–Hopf theorem.

Finally, we have arrived at the last step of the proof. We need to prove relation $(\#)$. If we recall that a surface can be deformed freely by using continuous deformation, we can create two vector fields (I) and (II) on S_g as shown in Figure 2.21.

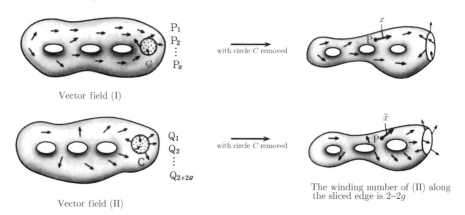

FIGURE 2.21

All critical points are assumed to lie inside the circle C. The pictures on the right-hand side of Figure 2.21 represent the surface S_g with the inside of the circle C removed. To make the explanation clearer, the interior of the circular disk is removed from S_g.

Now we examine the difference of directions of vectors in the vector fields (I) and (II) at each point of the surface $S_g - \widetilde{C}$ (where $\widetilde{C}$ represents the interior of the circle C). Let $\tilde{x}$ be the vector of the vector field (I) at the point P in $S_g - \widetilde{C}$, and x the vector of the vector field (II) at the same point. Notice that both x and $\tilde{x}$ are nonzero vectors. Consider the angle between x and $\tilde{x}$ at each point of $S_g - \widetilde{C}$ (see Figure 2.22). We denote this angle by

$$\theta(P) = \angle x\tilde{x}.$$

If we measure the angle between the vectors x and $\tilde{x}$ on the unit circle from the x-axis, as shown in Figure 2.22, we can determine the point representing $\theta(P)$ on the unit circle for each point P. In this way, with each point P on the surface $S_g - \widetilde{C}$, we can associate the point $\theta(P)$ on the unit circle. This correspondence is written as follows:

$$\theta\colon S_g - \widetilde{C} \to \text{unit circle}.$$

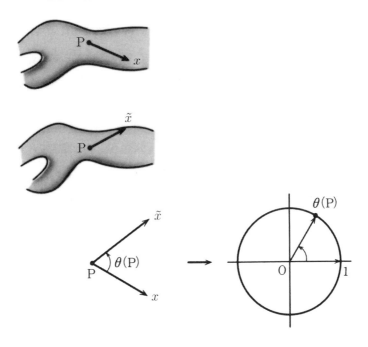

FIGURE 2.22

If the original point P in $S_g - \widetilde{C}$ moves slightly, then the vectors x and $\tilde{x}$ at P change slightly; thus $\theta(P)$ changes continuously as P changes.

Now let us slice the surface $S_g - \widetilde{C}$ from the left. (We illustrate this in Figure 2.23.) In this figure P_0 is a point on the surface; let $\theta(P_0)$ be the corresponding point on the unit circle. Consider the circle that appears

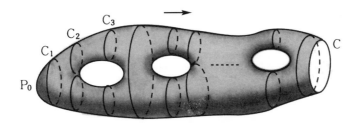

FIGURE 2.23

on the sliced edge of the surface slightly to the right of the point P_0. If P moves along the sliced edge C_1, then the winding number associated with $\theta(P)$ can be defined. Since $\theta(P)$ is continuous, the range of movement of $\theta(P)$ on C_1 is always near $\theta(P_0)$. Hence, as P moves completely around the sliced edge C_1, the point $\theta(P)$ cannot perform a complete revolution around the unit circle. Therefore, we can say that the winding number of θ along C_1 is 0. Even if we move the sliced edge to the right until it reaches the position C_2, property (a) of the winding number shows that the winding number of θ along C_2 is 0. In the case where C_2 splits into two circles, the winding number is still 0. If θ had the winding number n around the upper circle, then the winding number of θ around the lower circle would be $-n$. Therefore, by property (b) of the winding number the total winding number is 0. By property (a) of the winding number, the same happens at C_3 and so on.

As you can see, when the sliced edge of a surface $S_g - \widetilde{C}$ moves to the right, the winding number of θ around the the sliced edge is always 0. When we reach the farthest right edge C, the winding number of θ is still 0. This suggests that the difference between the winding number of the vector field (I) and that of the vector field (II) along C is 0. This means that relation ($\#$) is valid. Thus, we proved the Poincaré–Hopf theorem.

If we combine the proof of the Poincaré–Hopf theorem and the idea of the vector field constructed by barycentric subdivision of a triangulation of a surface (which was explained at the end of Section 2.3), we obtain another proof of the theorem that the Euler characteristic of a surface S_g does not depend on triangulation.

CHALLENGE 9. Verify that the Euler characteristic of a surface S_g does not depend on triangulation.

♣ In the Preface, we mentioned that some of our students had made suggestions to improve the clarity of the above proof. One of the suggestions was to slice a surface $S_g - \widetilde{C}$ into sections, and consider the winding number along the cross-sections. Our original explanation was not so clear to the students. Therefore, we used the suggested idea. If you have any suggestions that might help to clarify some of the explanations, or simplify the proof, please let us know by writing to us.

Curvature of a Surface
and the Euler Characteristic

In the third lecture we discuss the Gauss–Bonnet theorem, which shows the relationship between the curvature of a surface and its Euler characteristic. As we have said in the beginning of this book, the Gauss–Bonnet theorem provides a bridge between two different types of geometry—differential geometry and topology. The theorem was established in the 19th century and served as a foundation for the development of many important theorems of the 20th century mathematics. One might say that studying this lecture is like navigating a ship towards the 20th century mathematics.

3.1. Ellipses, Parabolas, and Hyperbolas

The Gauss–Bonnet theorem surrounds us in everyday life. If we look at things around us—the furniture in a room, dishes we use, art sculptures displayed in a museum, ocean waves pounding the beach, or even distant mountains—we notice that each of them has a curved surface. Sur*face*, as the word suggests, is the curved *face* of an object (however, not all surfaces are curved; a flat plane is also a surface). There are various types of surfaces with many different contours. How should we distinguish one surface from another? What point of view is necessary to accomplish this?

In preparation for establishing this point of view, let us introduce an interesting observation common to ellipses, parabolas, and hyperbolas. While this observation may seem unrelated to the subject at this point, it will prove to be relevant later.

When you hear the words "ellipse, parabola, and hyperbola," you will probably associate the words with the following equations:

Ellipse:
$$\frac{x^2}{a^2} + \frac{y^2}{b^2} = 1 \quad (a \geq b > 0),$$

Parabola:
$$y = ax^2 \quad (a \neq 0),$$

Hyperbola:
$$\frac{x^2}{a^2} - \frac{y^2}{b^2} = 1 \quad (a, b > 0).$$

These equations are written in the simplest form to express the corresponding curves mathematically. If we draw their graphs on a coordinate plane, we obtain the following three well-known types of curves:

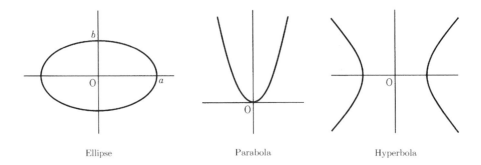

Ellipse   Parabola   Hyperbola

FIGURE 3.1

These three curves have absolutely different shapes. The ellipse has the shape of a slightly flattened circle. The region bounded by this curve is a finite region, and one can trace around an ellipse many times. The parabola is something like the orbit of a comet traveling from a distant point of the universe, reaching the vertex and continuing off again into space. In the figure above, the vertex is located at the origin of the coordinate plane. The hyperbola can best be compared to two different comets approaching each other from two different points of the universe. Once they reach the vertices of the hyperbola, they change direction and travel away from one another into space.

These three curves are called conic sections. At the second century B.C., their properties were explained in Appolonius' eight-volume work "*Conic Sections*". Even though these three curves are very different from one another, Appolonius saw that they are related through a double cone, as follows.

Imagine a straight line passing through the origin in the coordinate space and not coinciding with any coordinate axis. Call this line a generatrix. Rotating the generatrix around the vertical axis, we obtain two cones which are symmetric with respect to the origin (see Figure 3.2). In the case of Figure 3.2, the z-axis is the axis of the cone, but one could just as easily use the x-axis. Now imagine slicing the cone with a plane and focusing on the curves that appear at the section. This is best seen in Figure 3.2.

In case I the cone is sliced across by a plane which is perpendicular to the axis of the cone. The intersection curve is a circle. In case II, the cutting plane is tilted slightly and the intersection curve is an ellipse. In case III, the plane is tilted further to such a degree that it becomes parallel to the generatrix. The cone is cut by the plane, and the intersection curve is a parabola. In case IV, the plane is tilted even further, slicing through both upper and lower cones, and the intersection curve is a hyperbola. As

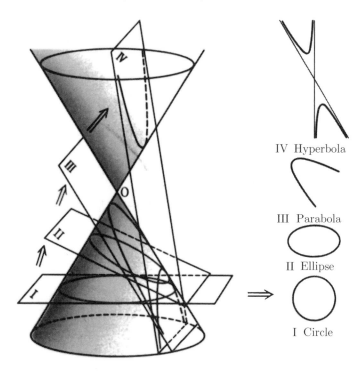

FIGURE 3.2

the position of the plane changes continuously, three different curves appear
naturally in the following order:

$$\text{Ellipse} \to \text{Parabola} \to \text{Hyperbola}.$$

Figure 3.3 gives a flow chart of how the intersections of a double cone and
the various positions of a tilted plane create conic sections. When the slope
of the plane in case II increases, the major (i.e., the longest) axis of the
ellipse increases and eventually changes the shape of the curve from finite
to infinite. When the slope of the plane is parallel to a generatrix as in
case III, we have a parabola. Moving along the chart further, we see that
the parabola splits into two branches and the intersection curve becomes
a hyperbola (case IV). Now we see how Appolonius observed the transi-
tion from finiteness to infinity using the double cone to produce these three
different types of curves.

This analysis leads to a new point of view related to the main subject of
the lecture (which we will discuss later). When we tilt the plane moderately,
as in case II, the ellipse changes its shape slightly but still remains an ellipse.
The same can be said about a hyperbola—when we make a slight change
to the slope of the plane, the shape of the hyperbola also changes slightly
but the curve still remains a hyperbola. We can say that an ellipse and a
hyperbola are stable. But the parabola is different if we look at it from the
point of view of conic sections. When we slightly tilt the plane, as shown

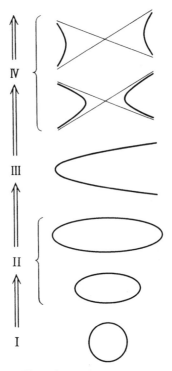

Flow chart of intersections
of a double cone and a
plane from various positions

FIGURE 3.3

in case III, the parabola immediately changes into either an ellipse or a hyperbola. This shows that the parabola is not stable.

The parabola is situated between two curves—the ellipse, which is finite, and the hyperbola, which is infinite. The parabola is a curve which appears on the "border" of finiteness and infinity.

3.2. Tangent Planes and Curved Surfaces

Previously, we discussed the finite nature of ellipses and the infinite nature of hyperbolas. We also discussed the fact that parabolas are curves that appear on the border of the finite and the infinite. Given all that, we can also say that the stability of ellipses and hyperbolas and the instability of parabolas illustrate how a surface is curved. Thus, Appolonius' discovery of the relationships between conic sections served as a ray of light—albeit from a barely opened door—for mathematicians who followed him (the door was opened further with the development of differential geometry and the technique of using tangent planes to study how a surface is curved).

We shall discuss tangent planes first. Let S be a surface and P a point on S. Imagine S as the surface of an apple. Now cut this apple in one direction through P. We obtain a space curve on the perimeter of the section of the apple. Now, keeping the two pieces together, cut the apple in a different direction through the same point. We obtain a different space curve on the perimeter of the section. Now imagine both of these cuts on the apple at the same time; we will have two different space curves passing through P (see Figure 3.4).

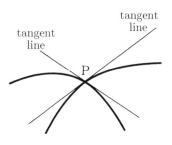

FIGURE 3.4

If we draw a tangent line to each of these two curves at the point P, we will obtain two straight lines passing through P. These two lines determine exactly one plane which will sit atop them like a window lying on its cross-piece. This plane passes through P and is called the **tangent plane** to S at P.

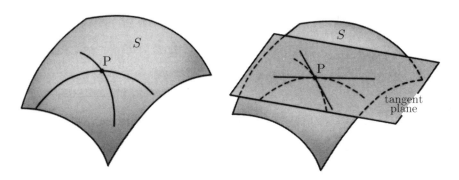

FIGURE 3.5

Some readers might wonder whether there might be another tangent plane at P if the surface were to be sliced in different directions. Such a question is quite reasonable. In fact, we need not worry about the existence of another tangent plane at P if the surface is smoothly curved. However, to avoid confusion, we consider only the case where for each point P on S, there is exactly one tangent plane at P.

As a way to observe the curving of the surface we use a method in which we slightly shift the tangent plane at P parallel to itself up and down and look at the intersection curve of this shifted plane and the surface.

First, we consider a sphere. At any point P on the sphere, if we slightly shift the tangent plane at P parallel to itself, we obtain a circle as the intersection curve of the plane and the sphere. Stating this in a more precise manner, we can say that if we move the tangent plane at the north pole upward, the plane does not intersect the sphere (in this case the intersection is referred to as the empty set, and the notation for the empty set is $\varnothing$). At the north pole, the tangent plane intersects the sphere at a single point P. If we shift the plane lower, the curve of intersection will be a circle.

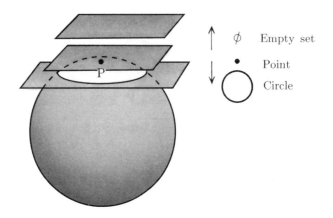

FIGURE 3.6

If we apply the above procedure to a convex region of a surface, we obtain similar results. Let P lie in the convex region of the surface. By shifting the tangent plane at P parallel to itself we will obtain the intersection curve that is similar to an ellipse. (In Figure 3.7, the caption reads "ellipse," but more precisely it should be "a curve similar to an ellipse.")

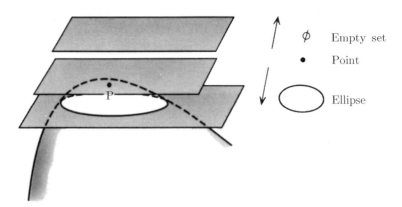

FIGURE 3.7

The region of the surface that shows such a curving as described above is said to have **elliptic curving**, and the point P is called an **elliptic point**.

On the other hand, a different type of curving occurs at a point P on the surface of an object shaped like a saddle, as in Figure 3.8. Such a point appeared earlier in the book when we discussed the index of a vector field on a surface. An example of such a point is a singular point of a hyperbolic

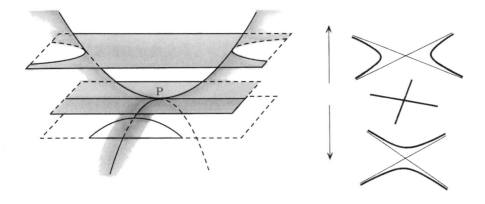

FIGURE 3.8

flow which has the index -1. There we observed the flow of rainwater at the pass of a mountain. The tangent plane at P cuts through the surface and two distinct straight lines appear at the intersection. If we slightly shift this tangent plane upward parallel to itself, the curve of intersection changes into a hyperbola. More precisely, we obtain a curve resembling a hyperbola. If we now slightly shift this tangent plane downward, the resulting intersection curve is another hyperbola (Figure 3.8). A person standing at the point P will feel as if he is in a valley between two peaks when looking up. But if one were to look down, one will have the feeling of being on the summit. This situation is represented by hyperbolas.

The region of the surface which shows this type of curving is said to have **hyperbolic curving**. We call the point P a **hyperbolic point**.

Every point on the surface of a sphere is an elliptic point and the sphere shows elliptic curving everywhere. The surface of a rugby ball or the surface of an egg are other common examples of elliptic curving. On the other hand, every point on the surface of a so-called hyperboloid of one sheet (Figure 3.9) is a hyperbolic point, and every region of the surface shows hyperbolic curving.

If a point P on a surface is an elliptic point, then every point near P is also an elliptic point. If a region of the surface shows elliptic curving, then even if that region is slightly deformed, it will still show elliptic curving. In other words, if a convex region of the surface is deformed slightly, it remains convex. Therefore, we can say that elliptic curving is stable.

Similar arguments show that hyperbolic curving is also stable. Imagine a saddle made of clay. Even if we stretch or squash the saddle slightly, its basic shape will remain the same. Now, if we think of this as an analogy to

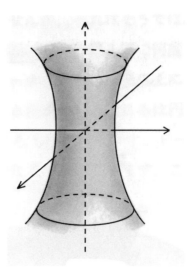

FIGURE 3.9

conic sections, we might imagine that there exists such a thing as parabolic curving, which must be situated between hyperbolic curving and elliptic curving.

Consider a ball made of clay. Every point on the surface of the ball is an elliptic point. But if we press it with our fingers and make an indentation, then hyperbolic points appear on the indented region of the surface. In this process of changing from convexity to concavity, parabolic curving appears at the moment where convexity changes into concavity. If we move even slightly away from this moment, parabolic curving immediately changes to elliptic or hyperbolic curving. Thus, we can say that parabolic curving is unstable.

Note that a point on the surface that is neither elliptic nor hyperbolic is called a **parabolic point**, and at parabolic points the surface has **parabolic curving**. Parabolic curving does not necessarily mean that a parabola appears as the intersection curve when the tangent plane at the parabolic point on the surface is slightly shifted parallel to itself.

A typical example of a surface which has parabolic curving is a plane. As we know, the tangent plane at point P on the plane is the plane itself. If we shift this tangent plane up and down parallel to itself there will be no intersection (Figure 3.10).

Also, every point on the surface of a cylinder shows parabolic curving. In this case, the intersection of the tangent plane and the surface is a line as in Figure 3.10. However, if we slightly shift the tangent plane upward, then the intersection immediately becomes the empty set, and if we slightly shift it downward, the intersection will be two parallel lines.

If we now look at the surface of a cone keeping the aforesaid in mind, we will see that a parabola appears at the intersection of the shifted tangent

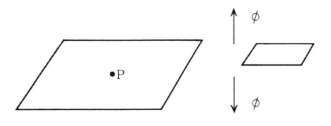

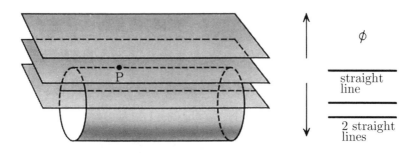

FIGURE 3.10

plane and the cone. Thus, we can conclude that every point on the cone is parabolic (Figure 3.11).

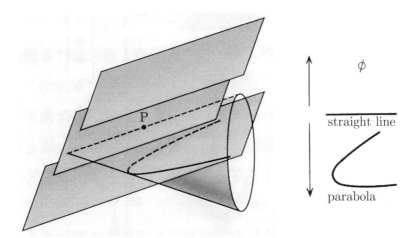

FIGURE 3.11

Some readers might have conjectured that a surface which changes its curving in the following succession:

(∗)   Elliptic → Parabolic → Hyperbolic

is a cone. However, this is not the case. A simple example of such a surface is the surface of a doughnut (a torus). When a doughnut is placed on a flat desk, and a piece of paper is placed on top of it, the intersection of the doughnut with the paper is a circle, and the intersection of the doughnut with the desk is also a circle. On the surface of the doughnut, the points on these two (upper and lower) circles are parabolic because if we move away from these points, the curve of intersection immediately changes. The outer region of the surface of the doughnut between these two circles consists of elliptic points, and the inner region of the surface of the doughnut between the two circles consists of hyperbolic points. The change of the curving as in (∗) is observed clearly on the surface of the doughnut.

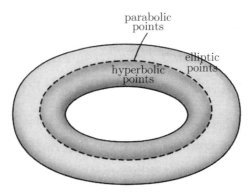

FIGURE 3.12

CHALLENGE 10. Verify the above assertion by shifting the tangent plane up and down and observing the intersection curve.

3.3. Curvature and the Gauss Map

As we described before, the curving of a point on a surface is divided roughly into three categories: elliptic, hyperbolic, and parabolic. Elliptic curving itself has a wide range of shapes from very sharp curving (almost as sharp as the tip of an icicle) to very flat curving (almost as flat as a plane). Hyperbolic curving has a wide range of shapes as well.

Is there any way to measure the curving quantitatively? In other words, is there a real number that shows the curving at each point of the surface (whether those points are elliptic, hyperbolic or parabolic), and the degree thereof?

Indeed, it is possible to find such a number using a measurement method called the **Gaussian curvature**. The Gaussian curvature is a quantity which shows the extent to which the surface curves, and it is given by the real number corresponding to the degree of curvature (Figure 3.13).

According to Figure 3.13, the Gaussian curvature is positive when the curving of the surface is elliptic. The sharper the elliptic curving, the greater

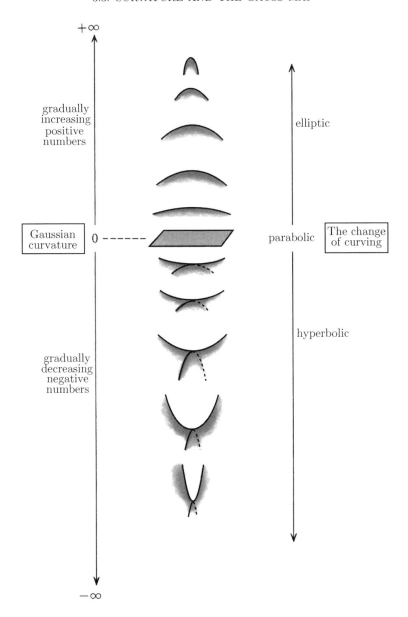

$+\infty$

gradually
increasing
positive
numbers

elliptic

| Gaussian curvature | 0 - - - - - - |

parabolic

| The change of curving |

hyperbolic

gradually
decreasing
negative
numbers

$-\infty$

FIGURE 3.13

the Gaussian curvature, and if the elliptic curving is small, then so is the
Gaussian curvature. If the surface is parabolic like a plane, then the Gaussian curvature is zero. If the surface starts to show hyperbolic curving such
as a saddle, then the Gaussian curvature becomes negative, and the sharper
the hyperbolic curving of the surface, the smaller the Gaussian curvature.
As we see in Figure 3.13, each point on a real line represents the curvature
of a surface measured by the Gaussian curvature.

So how shall we define the Gaussian curvature? Actually, the Gaussian curvature can be expressed by a complicated formula that involves equations describing a surface. However, Gauss himself showed in his famous historical essay that the Gaussian curvature can be defined using a more geometric approach. Let us now discuss this principle.

> ♣ Gauss was one of the most prominent German mathematicians of the nineteenth century. His creative achievements covered all areas of mathematics, and even extended into mathematical physics. It might be said that Gauss alone had subjugated various peaks of mathematics that rose high above the clouds. Gauss considered number theory to be the queen of mathematics and the central theme of his research. His introduction of the Gaussian curvature, though atypical of his strictly integer-oriented research, was a stunning and unprecedented achievement marking the beginning of a new era in mathematics. His work, along with the work of another great mathematician, Riemann, facilitated the birth of a new branch of mathematics called differential geometry. Gauss' essay was published in 1827 under the title "The General Theory of Surfaces."

The fundamental idea behind the introduction of the Gaussian curvature is the Gauss map, which is a mapping from a surface to the unit sphere.

> ♣ We will briefly explain functions and mappings. If, for each real number x, there is an associated real number y, then y is called a function of x and is denoted by
>
> $$y = f(x).$$
>
> If for each point P in space, a point Q in space is determined, then this correspondence is also a function but it is specifically called a mapping and is denoted by
>
> $$Q = f(P).$$
>
> Using the word "mapping" may help the reader to understand and visualize the concept of a function.

Consider a sphere with radius one unit. We call such a sphere the **unit sphere**. To visualize this sphere, we need to know its location. To avoid ambiguity, consider an xyz-coordinate space. Now consider the sphere of radius 1 centered at the origin of this coordinate space, as in Figure 3.14. In this way, the definition of the unit sphere is clearer and easier to visualize.

To introduce the Gauss map, Gauss focused on a normal vector at P to the surface S. A **normal vector** at P is a vector which is orthogonal to the tangent plane at P. A **unit normal vector** is a normal vector whose length is one unit.

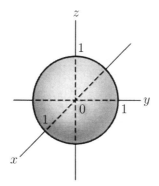

FIGURE 3.14

At each point on the surface S, there are exactly two unit normal vectors. We will not go into detail about which one to choose here, but let us take a vector coming outward from the closed surface to be the unit normal vector (Figure 3.15). In this way we can imagine unit normal vectors poking out everywhere from a closed surface like the back of a porcupine.

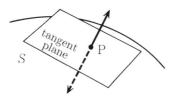

FIGURE 3.15. On each side of the tangent plane, there is one normal unit vector at P.

The unit normal vector at the point P will be denoted by $\vec{n}_P$. The initial point of the vector $\vec{n}_P$ is P. Now apply a parallel translation to this vector such that the beginning of the vector, which was the point P, will coincide with the origin of the xyz-coordinate space. After this parallel translation, the endpoint of the vector $\vec{n}_P$ lies on the unit sphere.

This process determines, for each point on the surface S, a point on the unit sphere—that is, the endpoint of the corresponding vector. This point on the unit sphere is denoted by $g(P)$. As the point P moves on the surface S, the point $g(P)$ moves on the unit sphere. A mapping from the surface S to the unit sphere,

$$g\colon S \to \text{unit sphere},$$

is thus defined. We call this mapping the **Gauss map**. If you find the above explanation unclear, studying Figure 3.16 may help.

Let us explain the Gauss map using the three examples illustrated in Figure 3.17.

Example (I) shows the Gauss map from a sphere of radius R to the unit sphere. Two points P and Q on the sphere are mapped to the points $g(P)$

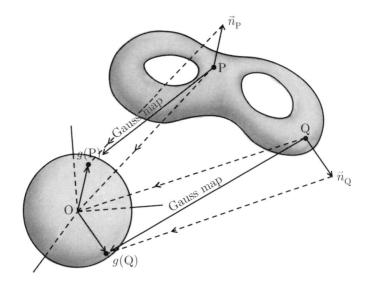

FIGURE 3.16

and $g(Q)$ on the unit sphere where normal vectors at $g(P)$ and $g(Q)$ have the same directions as those at P and Q, respectively. In this case, the Gauss map is a contraction (or a dilation) of the sphere of radius R to the unit sphere. This means that we have a similar transformation with the similarity ratio $1/R$.

Example (II) shows the Gauss map from a surface shaped like a lemon. A section of the large upper region of the lemon is mapped (contracted) to the smaller region near the north pole of the unit sphere. On the other hand, the sharply curved part of the lemon (as opposed to the stem end) is mapped (expanded) to the region almost as large as half of the total area of the unit sphere. By observing these figures, we can see that the Gauss map shows the degree of curving of the surface by contracting or expanding the region of the surface mapped to the unit sphere. This point of view is the one on which Gauss focused his attention.

In example (II), we can see how the part of the surface with elliptic curving is mapped by the Gauss map.

Example (III) shows us how the part of the surface with hyperbolic curving is mapped by the Gauss map. If we look carefully, we see that the left and right sides of the region divided by the line through the point P are mapped to the opposite sides of the line which passes through the point $g(P)$ on the unit sphere. Also, a small region containing the point P is mapped inside out to the unit sphere. We can recognize a hyperbolic curving immediately by this characteristic property of reversing (inside out) under the Gauss map, as shown in Figure 3.17(III).

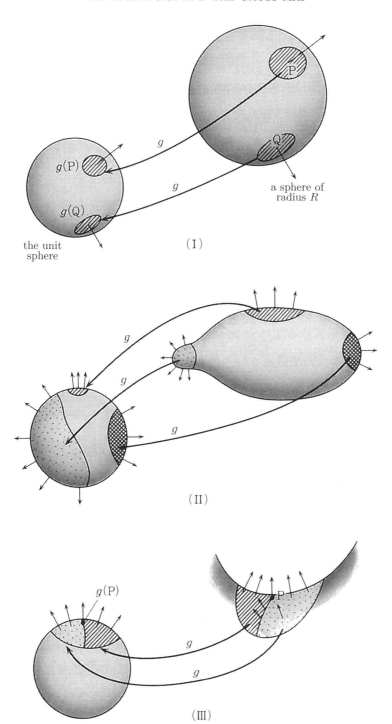

the unit
sphere

(I)

a sphere of
radius R

(II)

(III)

FIGURE 3.17

3.4. Gaussian Curvatures

From Figure 3.17(II), one can conclude that the curving of a region around a point P on a smooth surface is represented by the ratio of the area of the region σ containing the point P, and the area of the mapped region $g(\sigma)$. Thus, we are ready to define the Gaussian curvature. First let us decide the order of the ratio we will use. Should we set up the ratio as "Area of $g(\sigma)$: Area of σ" or "Area of σ : Area of $g(\sigma)$" ? We should set up the ratio to reflect the principle of "the greater the curvature, the greater the curving." If we look at Figure 3.17(II), we will see that as the curving of σ increases, the area of the mapped region $g(\sigma)$ increases. Therefore, the ratio

$$(1) \qquad \frac{\text{Area of } g(\sigma)}{\text{Area of } \sigma}$$

should be the starting point for discussing the Gaussian curvature.

Observe that expression (1) does not show the curving of a surface at the point P. Rather, it shows the *average* value to indicate the curving of the region on the surface S containing P. As this region σ containing P shrinks smaller and smaller, the average value approaches the value that represents the degree of curving at the point P. At this time, the reader might protest that it is meaningless to talk about the curving at a point; also the curving of a region containing the point P gives us only a rough idea about the surface. Indeed, we have a complicated, paradoxical phenomenon. When mathematicians of the eighteenth century began to study a surface as an object of geometry, this phenomenon was one of many difficulties they encountered. We now take a moment to examine this.

We can overcome this difficulty if we apply the concept of the limit, as used in calculus, to the surface. Consider the limit of expression (1) and find

$$(2) \qquad \lim_{\sigma \to P} \frac{\text{Area of } g(\sigma)}{\text{Area of } \sigma}.$$

In this expression, the notation $\lim_{\sigma \to P}$ means that if the region σ containing P shrinks smaller and smaller to the point P, we can obtain a real number as the limit of expression (1). However, if the region on the surface near point P is curved up and down irregularly, is it still possible to obtain an exact number as the limit of expression (1)? The answer is yes, because if the surface is smooth, one can guarantee that the limit of expression (1) always exists. Thus, we always obtain exactly one number for expression (2).

For us to regard expression (2) as the curvature of the surface at the point P is a bit premature. We first need to examine the situation where the curving of the surface is hyperbolic as in Figure 3.17(III), keeping in mind the reversing phenomenon of the Gauss map. Next, we need to reflect this reversing phenomenon in the concept of the curvature. To accomplish this, we start by drawing a circle centered at P in the positive direction (counterclockwise) inside the region σ. When this circle is mapped by a

Gauss map, the image circle will be inside out, the reversal of right and left will occur, and the direction of the image circle around the point $g(P)$ will be negative. In such a case we will assign a negative sign to the area of $g(\sigma)$. This area is called the **signed area of** $g(\sigma)$.

At this time, the Gaussian curvature $K(P)$ of a surface S at a point P can be defined as follows:

$$(3) \qquad\qquad K(P) = \lim_{\sigma \to P} \frac{\text{signed area of } g(\sigma)}{\text{area of } \sigma}.$$

From the above discussion we can determine the following facts relating to Gaussian curvature in Figure 3.17(I), (II), and (III).

Example (I): The Gaussian curvature of a sphere of radius R is constant at each point of the sphere. It is given by

$$K(P) = \frac{1}{R^2}.$$

We can make this conclusion because if a sphere of radius R is mapped to the unit sphere by the similar transformation with the similarity ratio $1/R$, then the ratio of the corresponding areas is the square of the similarity ratio, which is $1/R^2$.

Example (II): In this case we observe that the reversing phenomenon does not occur at a point on the surface with elliptic curving; therefore, we have $K(P) > 0$. The greater the curving, the greater the number $K(P)$.

Example (III): In this case we observe that at a point on the surface with hyperbolic curving, the reversing phenomenon occurs and $K(P) < 0$.

Summarizing the above, we have

> The Gaussian curvature of a sphere of radius R is constant and equal to $K(P) = 1/R^2$.
> At an elliptic point P, $K(P) > 0$.
> At a hyperbolic point P, $K(P) < 0$.

If the curving at a point Q is greater than that at a point P, then

$$K(P) < K(Q)$$

whenever the curving is elliptic, and

$$K(P) > K(Q)$$

whenever the curving is hyperbolic (see Figure 3.18).

Now we ask what is the curvature at a point which shows parabolic curving? First keep in mind that parabolic curving is located between hyperbolic curving and elliptic curving. From this observation, we can infer that if a point P has parabolic curving, then

$$K(P) = 0.$$

To confirm this inference, let us examine three illustrations in Figure 3.19

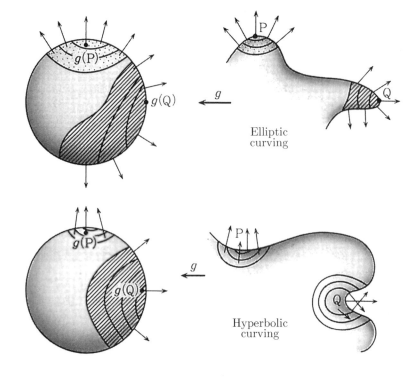

FIGURE 3.18

A plane, which always shows parabolic curving, is mapped to just a point on the unit sphere by the Gauss map. Therefore, the numerator of expression (3) is 0; hence, $K(P) = 0$. Cones and cylinders both show parabolic curving. They are mapped to a circular arc on the unit sphere; therefore

$$K(P) = 0.$$

We have now shown that the Gaussian curvature is a quantity which shows the degree of curving of a surface.

3.5. The Gauss–Bonnet Theorem

Finally, we discuss the Gauss–Bonnet theorem,[1] the main subject of Lecture 3. If we were to discuss this theorem in detail, we would need to establish the theory of integration over a surface. We need not do this since the reader must only understand the content of the Gauss–Bonnet theorem on a basic and intuitive level.

Speaking generally, the content of the Gauss-Bonnet theorem is stated as follows:

[1]The actual Gauss–Bonnet theorem states that if S is a compact orientable surface of class C^3, then $\int_S K(P) \, d\sigma = 2\pi \chi(S)$.

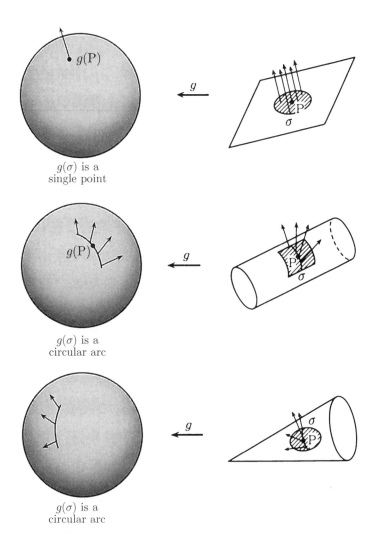

$g(\sigma)$ is a
single point

$g(\sigma)$ is a
circular arc

$g(\sigma)$ is a
circular arc

FIGURE 3.19

GAUSS–BONNET THEOREM. *The "total sum" of the Gaussian curvature K over a surface is equal to the Euler characteristic of the surface multiplied by 2π.*

The "total sum" is represented by the integral

$$(4) \qquad \int_S K(P)\, d\sigma.$$

To clarify things a little, let us discuss the "total sum." First, partition a closed surface S into sufficiently small regions σ_1, σ_2, ..., σ_n. We can assume that over each region σ_i the curving is moderate; hence, the Gaussian curvature of each region varies moderately, becoming almost constant. Next, take the following points: P_1 in the region σ_1, P_2 in the region σ_2, ..., P_n in the region σ_n. Finally, consider products of the value of the Gaussian curvature at the point P_i and the area of the region σ_i ($i = 1, 2, \ldots, n$):

$$K(P_1) \times (\text{area of } \sigma_1), \ K(P_2) \times (\text{area of } \sigma_2), \ \ldots, \ K(P_n) \times (\text{area of } \sigma_n).$$

By replacing the Gaussian curvature at each point in σ_1, σ_2, ..., σ_n with constant values $K(P_1)$, $K(P_2)$, ..., $K(P_n)$, respectively, we can regard each product as representing the average of the total sum of the Gaussian curvature of each partitioned region σ_i.

> ♣ If you find the words the "average of the total sum of the Gaussian curvature" confusing, simply observe only the mathematical expressions, and do not worry about the wording.

Now sum up the products as follows:

(5) $\quad K(P_1) \times (\text{area of } \sigma_1) + K(P_2) \times (\text{area of } \sigma_2) + \cdots + K(P_n) \times (\text{area of } \sigma_n).$

We will consider this sum to be an approximation of the total sum of the Gaussian curvature over the surface S. When the parts σ_1, σ_2, ..., σ_n are subdivided into smaller and smaller regions, expression (5) tends to a limit and this limit is expressed by the integral notation, as in (4). Now the Gauss–Bonnet theorem can be expressed as the equality

(6) $$\frac{1}{2\pi} \int_S K(P)\, d\sigma = \chi(S).$$

The Gauss–Bonnet theorem allows us to understand something very interesting. When we see children playing with Play-Doh® and observe them constantly changing the shape of the dough freely, or when we see a glass blower creating a vase by skillfully changing the shape of the glass by blowing air into it, we might think that the concavity and convexity of a surface can be changed without any restrictions. In fact, if we limit the workable region to a certain part of the surface, we can deform this part of the surface in a concave or convex manner. But when we dent a certain part of the surface in or out, then another part or parts have an opposite reaction. In other words, pushing a certain part might cause another area to lose its convexity and even to become dented inwards. This happens because the surface is closed.

If the surface is like a plane, which is infinitely spreading, or is like a polygon with a boundary, we can make the entire surface convex so that there is absolutely no indentation—like a glass dome. This cannot happen, however, in the case of a closed surface. The following phenomenon is unique to closed surfaces: if an indentation is made on a certain part of the surface, then a convexity will appear on another part of the surface. This is not such an obvious statement. Indeed, it seems too good to be true that such a phenomenon could be expressed mathematically.

The Gauss–Bonnet theorem gives a clear formulation to this phenomenon. According to the Gauss–Bonnet theorem, the way a surface is shaped with regard to convexity and concavity is completely controlled by the Euler characteristic. The Euler characteristic which appears on the right-hand side of equation (6) remains constant when the surface is deformed continuously. If the surface is deformed and some part of it starts having a greater Gaussian curvature, then the curvature of some other parts of the surface starts decreasing. In this way, the balance of concavity and convexity is maintained, and the left-hand side of equation (6), which is the total Gaussian curvature, remains constant.

The deeper we understand the content of this theorem, the more we realize its enormous depth, and the more inspiring it becomes.

Let us give the proof of this theorem in a more intuitive form.

First, we will see that equation (6) holds for a sphere of radius R. In this case, the Gaussian curvature is constant at each point of the sphere, and it is equal to $1/R^2$. The surface area of this sphere is $4\pi R^2$. Therefore, by referring to the definition of the integral, we see that the left-hand side of equation (6) is equal to

$$
\frac{1}{2\pi} \int_S K(P)\, d\sigma = \frac{1}{2\pi} \times \frac{1}{R^2} \times \int_S d\sigma
$$
$$
\left(\text{here } \int_S d\sigma \text{ is the surface area of a sphere} \right)
$$
$$
= \frac{1}{2\pi} \times \frac{1}{R^2} \times 4\pi R^2
$$
$$
= 2.
$$

Recall from Lecture 1 that the notation for a sphere is S_0 and the Euler characteristic of S_0 is 2. Hence, for a sphere of radius R, equation (6) holds.

How shall we then prove the case where the surface looks like a sphere but has many concavities and convexities? More generally, how shall we prove the case where the surface is a closed surface of genus g?

Let us look at the Gauss map closely one more time.

We previously gave the definition of the Gauss map $g\colon S \to$ unit sphere as follows. Given a closed surface of genus g in three-dimensional space, at each point P on the surface, take the unit normal vector $\vec{n}_P$. Then apply a parallel translation to this unit vector so that the initial point of $\vec{n}_P$ moves to the origin of three-dimensional space (the center of the unit sphere). Then the endpoint $g(P)$ of the unit normal vector lies on the unit sphere. The Gauss map is a correspondence that associates to each P on S the endpoint $g(P)$ on the unit sphere. We say that $g(P)$ is the image of P under the Gauss map. The four basic properties of the Gauss map are as follows:

(1) Assume that P is an elliptic point on a surface S. The unit normal vectors at the points near P all have different directions from that of $\vec{n}_P$

with $\vec{n}_P$ in the center (see Figure 3.20). Therefore, if we consider a sufficiently small region σ containing the point P, σ will be mapped to a region
on the unit sphere containing the point $g(P)$ without any holes. Of course,
the image region can be larger or smaller according to the curvature.

If we focus on a small region containing the elliptic point, the above
fact can be understood by observing the roundness of the surface just like
a sphere (see Figure 3.20). If we first imagine the picture in Figure 3.20
to be that of the opening of flower petals around the center of the flower
represented by $\vec{n}_P$ and we next translate the tips of the petals to the unit
sphere, then the mapping of the points of the small region on the surface to
the unit sphere is the Gauss map.

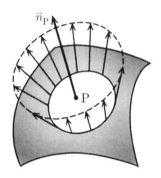

FIGURE 3.20

(2) Assume that P is a hyperbolic point. Again we observe that the unit
normal vectors at points near P all have different directions from that of $\vec{n}_P$
with $\vec{n}_p$ in the center. However, in this case the directions of the varying
vectors are reversed from right to left when compared to the elliptic case
(as shown in Figure 3.21). Hence, if a sufficiently small region containing P
is taken, then the image of this region on the unit sphere is inverted inside
out.

This case might be more difficult to vizulaize than the case of elliptic
points. Careful observation of how the direction of normal vectors vary in
Figure 3.21 will help you to understand this concept.

(3) Assume that P is a parabolic point. In this case, the directions of the
unit normal vectors at the points near P could all be the same as that of $\vec{n}_P$.
Even if they are different, the change in directions is restricted. Therefore,
if a sufficiently small region σ containing P is mapped to the unit sphere
by the Gauss map, σ might shrink to a point or to a curved segment on the
unit sphere and the image $g(\sigma)$ will not have any interior parts.

The above is true because when a point P on a surface S moves on S continuously, the corresponding unit normal vectors also change their directions
continuously, and the directions of the vectors will not change abruptly.

(4) The image of a closed surface S under the Gauss map to the unit
sphere wraps the entire unit sphere. This can be explained as follows. Let

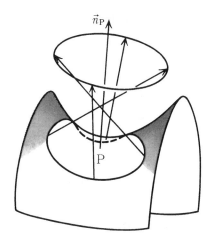

$\vec{n}_\text{P}$

P

FIGURE 3.21

Q be any point on the unit sphere. Now consider the tangent plane at Q. Move this tangent plane by a parallel translation sufficiently far away from the surface S, so that there will be no intersection between the plane and the surface S. Then, shift the plane back closer to the surface until it contacts the surface S at some point or points (it might be one point, several points, or infinitely many points). Let P be any one of these tangency points. Then, the point P must be an elliptic point or a parabolic point and $g(P) = Q$.

CHALLENGE 11. Verify the above fact by drawing the unit sphere, tangent planes, and a surface S.

To summarize (1)–(4) above, the following can be said to describe the Gauss map. First, imagine that the surface S is made of rubber. The mapping of a small region σ by the Gauss map to the unit sphere is analogous to cutting σ out of the surface S, then stretching it or shrinking it according to the curving (and/or reversing it), and then "gluing" it onto the unit sphere.

Simply put, S is cut into small pieces and then glued onto the unit sphere after some stretching, shrinking or reversing of each piece. The important thing in the gluing process is that each piece is glued in sequence so that neighboring pieces remain next to each other on the unit sphere. Also, there should be neither holes nor "tears" in any piece when we glue them to the unit sphere. The reader should keep in mind the first lecture, where we discussed how a surface may be constructed by gluing triangles together.

If we glue the pieces as above, the unit sphere will be completely covered by the glued pieces (property (4)), but this does not necessarily mean that the unit sphere is enwrapped only once. It could be covered many times with right-side out pieces or a combination of both inside-out pieces and right-side out pieces. Since breaking or puncturing the pieces is not allowed during the process of gluing (it is very important to remember this!), the unit sphere will be covered several times by the surface S via the Gauss map.

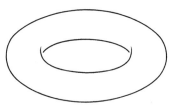

FIGURE 3.22

CHALLENGE 12. The surface of a doughnut in Figure 3.22 covers the unit sphere as follows: The regions of the surface composed by elliptic and parabolic points together completely cover the unit sphere via the Gauss map. The regions of the surface composed by hyperbolic points are turned inside out and they, together with parabolic points, also cover the unit sphere. Illustrate this fact with drawings.

Now that we understand the geometric properties of the Gauss map, let us explore the meaning of the following expression from the *left-hand* side of equation (6):

$$\int_S K(P)\, d\sigma.$$

This integral is given as the limit of the sum (5). However, by referring to the definition of the Gaussian curvature given by equation (3), $K(P_i)$ may be approximated as follows:

$$K(P_i) \approx \frac{\text{signed area of } g(\sigma_i)}{\text{area of } \sigma_i}.$$

If we substitute this approximation in the sum (5), we obtain the following:

(7) $$\int_S K(P)\, d\sigma = \lim \sum_i \text{signed area of } g(\sigma_i).$$

What does the right-hand side of this equation mean? Since the surface S is glued to the unit sphere, it is nothing but the total area of the glued pieces on the sphere. (We assign the negative sign to the area of the reversed part.)

As we know, the closed surface S covers the unit sphere several times via the Gauss map. In general, the number of layers is different for different points on the unit sphere. Since the region composed of parabolic points on the surface S will be shrunken to a single point or a curved segment once mapped to the unit sphere, we cannot define the number for counting the layers at these points. However, such points form a set of zero area; therefore, they are exceptional points.

Most of the points on the unit sphere (except those just mentioned) are covered by the surface S via repeated wrappings (in layers or sheets).

Therefore, we need to find a way to count the number of layers. As we know, if a small region of a surface S is mapped to the unit sphere inside out, the image of that region has a signed area (i.e., negative value). Keeping this fact in mind, we count the reversed covering to be -1. Hence, at every point on the unit sphere except for the images of parabolic points, the number of sheets covering the point is determined by an integer.

Now we formulate the following fact without actually proving it. The number of sheets covering the unit sphere as described above does not depend on the choice of the point on the unit sphere. This is so because the surface can be stretched or shrunken, but breaking or puncturing the surface is not allowed. We hope the readers will find this intuitively clear. Keeping this in mind, let us look at (7) again:

$$(8) \quad \int_S K(P)\, d\sigma = (\text{the number of sheets covering the unit sphere}) \times 4\pi,$$

where 4π is the surface area of the unit sphere.

Now let us look at a closed surface S of genus g as shown in Figure 3.23. There are $g + 1$ points on the surface S which will be mapped to the south

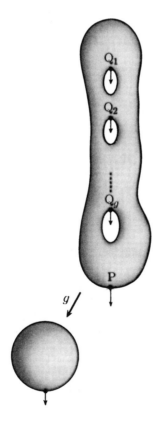

FIGURE 3.23

pole on the unit sphere, namely

$$Q_1, Q_2, \ldots, Q_g, P.$$

At the points Q_1, Q_2, ..., Q_g, the surface S shows hyperbolic curving. Therefore, we can conclude that the unit sphere is covered $1 - g$ times by the surface S via the Gauss map.

CHALLENGE 13. Show that if we consider the north pole instead of the south pole, the number of sheets covering the unit sphere via the Gauss map is also $1 - g$.

From equation (8), we obtain

$$\int_S K(P)\, d\sigma = (1 - g) \times 4\pi.$$

Therefore, the left-hand side of equation (6) is given by

$$\frac{1}{2\pi} \int_S K(P)\, d\sigma = \frac{1}{2\pi}(1 - g) \times 4\pi = 2 - 2g.$$

This is exactly the Euler characteristic of the surface S; that is,

$$\chi(S) = 2 - 2g.$$

Thus, we have proved the Gauss–Bonnet theorem for a particular surface S by showing that equation (6) is valid.

The general surface S_g of genus g is obtained by deforming the above surface S continuously. However, the continuous deformation illustrated in Figure 3.24 has a broader meaning: it allows the surface to "weave through" existing holes as shown in the figure.

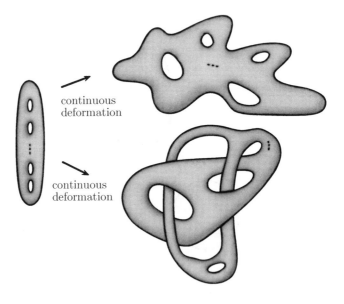

continuous
deformation

continuous
deformation

FIGURE 3.24

We now know that the Gauss map is a mapping that covers the unit sphere several times. We also know that the number of sheets covering the unit sphere does not change if the surface is slightly deformed (accordingly, the image of the surface on the unit sphere changes slightly). For example, if the unit sphere is covered by five layers, then this number cannot suddenly change to four. Even on the region where the surface passes through one of its own holes (as in Figure 3.24), the Gauss map is still defined properly and the number of layers near that "weaving" region does not change. Since the number of layers does not change after a slight deformation, if S is finally changed into S_g by continuous deformation, the number of layers remains the same. Referring to equation (8), we obtain the following formula:

$$\frac{1}{2\pi}\int_S K(P)\,d\sigma = \frac{1}{2\pi}\int_{S_g} K(P)\,d\sigma.$$

On the other hand,

$$\chi(S) = \chi(S_g) = 2 - 2g.$$

Consequently,

$$\frac{1}{2\pi}\int_{S_g} K(P)\,d\sigma = \chi(S_g).$$

Thus, we have proved the Gauss–Bonnet theorem.

3.6. Surfaces of Constant Curvature

Now that we have completed the proof of the Gauss–Bonnet theorem, let us discuss surfaces of constant curvature.

The curvature of the unit sphere is constant at each point on the sphere and is given by

$$K(P) = 1.$$

Therefore, a sphere of radius 1 is a closed **surface of constant positive curvature**. The curvature of a sphere of radius R is also constant at each point on the sphere and is given by $K(P) = 1/R^2$. So again, this is a closed surface of constant positive curvature.

In three-dimensional space, there is no closed surface whose curvature is 0 at all points. To demonstrate this, consider a closed surface S in xyz-space. Choose a point on S such that the distance from the origin to that point is the longest possible. Then the region around this point protrudes outward. Thus, the curvature at that point is positive (not zero). Therefore, a closed surface S in xyz-space cannot be a surface of constant curvature 0. (See Figure 3.25.)

If we do not restrict our discussion to closed surfaces, then we can have a plane or an infinitely stretching cylinder (which are not closed surfaces) as examples of surfaces of constant curvature 0. In these examples, every point on the surface is a parabolic point.

Imagine "pinching" the middle of an infinitely spreading cylinder, then stretching both sides out like two horns facing away from each other. The

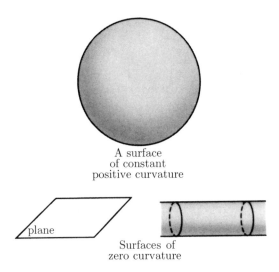

A surface
of constant
positive curvature

plane

Surfaces of
zero curvature

FIGURE 3.25

resulting figure is called a hyperboloid of one sheet. It is a surface whose curvature is negative at all points because all points are hyperbolic. A surface with constant negative curvature (that is, a surface which shows hyperbolic curving at each point) does not exist if we restrict our discussion to closed surfaces. But if we include unbounded (infinitely spreading) surfaces, then such a surface does exist. If a surface has negative curvature $K(P)$ at each point, it is called a **surface of constant negative curvature**. Unbounded surfaces of constant negative curvature in three-dimensional space are known to have singular points (nonsmooth points). Three examples are shown in Figure 3.26.

Negative or positive curvature indicates hyperbolic or elliptic curving of the surface, respectively. In the elliptic case, the surface is finite; in the hyperbolic case, it is infinite. This seems to reflect the essential nature of ellipses and hyperbolas as conic sections. The words elliptic and hyperbolic appear repeatedly in many branches of mathematics. Even in present day mathematics, when we look at these surfaces, we adopt the point of view which was first established by the work of Appolonius.

There is a "surface with boundary" whose curvature is constant and negative. We can obtain it as a surface of revolution and it is called the **pseudo-sphere**. Constructing this surface is intuitive and simple; once one hears about it, it is not easily forgotten. Imagine a dog walking along the x-axis in the positive direction dragging a leash of length 1 with a weight attached to the end of the leash. If the dog starts at the origin with the weight lying on the y-axis at length 1 away, then the path of the weight at the end of the leash is a curve on the plane. The curve obtained in this manner is called the **tractrix**. The surface of revolution generated by revolving the tractrix about the x-axis is called the **pseudo-sphere** and the

FIGURE 3.26

curvature of this surface is -1 at all point with $x > 0$. This pseudo-sphere is a surface of constant negative curvature.

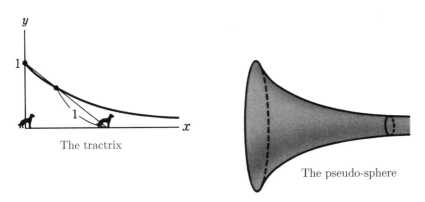

The tractrix

The pseudo-sphere

FIGURE 3.27

Until now we have assumed that closed surfaces have a real form that we can actually see with our eyes. That is, these surfaces appear to us as "real" figures in three-dimensional space in which we live. In this setting, the only closed surface of constant curvature is a sphere, and its curvature is positive. A closed surface of constant negative or zero curvature does not exist in three-dimensional space. When we say that it does not exist, we are thinking of the existence of figures in terms of three-dimensional space.

Now let us change our point of view. Instead of visualizing a figure of a doughnut in three-dimensional space, consider the surface of a doughnut obtained by identifying opposite sides of a rectangle $ABCD$ as we have done several times before. Consider the rectangle $ABCD$ as a tile placed on a plane (a floor). Now place eight other tiles around $ABCD$ as in Figure 3.28. Pick a point Q on the tile $ABCD$ and copy that point on the eight other tiles. Let Q_1, Q_2, $\ldots$, Q_8 represent the corresponding points on the eight other tiles.

Keeping this in mind, we introduce the distance between two points P and Q on the rectangle $ABCD$ as follows. Let the distance between P and Q be the length of the shortest of the segments PQ, PQ_1, PQ_2, $\ldots$, PQ_8. To better understand this definition, look at Figure 3.28.

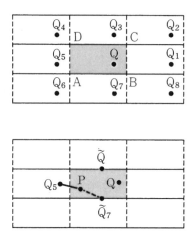

FIGURE 3.28. The "length" of PQ equals the length of PQ_5. The "length" of $P\widetilde{Q}$ equals the length of $P\widetilde{Q}_7$.

This distance gives us a new way to measure the length on the surface of a doughnut. Since the distance between two points on the plane is transferred onto the surface of the doughnut, it is intuitively clear that the curvature of this doughnut is zero. In fact, from the point of view of differential geometry, the surface of a doughnut with the above length has curvature zero, and such a surface is called a **flat torus**.

A flat torus has curvature zero at all points; therefore, it is a surface of constant curvature. However, without changing the length of the sides of $ABCD$ we cannot glue the opposite sides of $ABCD$ in three-dimensional space. The torus we are used to seeing is a curved surface where the segment which yields the shortest distance between two points is not a straight segment. If we cut open the torus along this segment, then we find the segment to be curved unlike the sides of the rectangle $ABCD$.

A flat torus does not have a shape in three-dimensional space, so in this sense it is imaginary.

However, if we eliminate the restriction of considering only three-dimensional space and consider higher-dimensional spaces instead (which we will discuss in Lecture 1 of Chapter 2), we find that imaginary figures take shape and a flat torus can be realized as a closed surface!

The same is true for a closed surface of genus g. In this case the Euler characteristic is negative. Therefore, according to the Gauss–Bonnet theorem, the average total Gaussian curvature is negative. But as we mentioned before, in three-dimensional space, constant negative curvature at each point of the surface cannot be realized. Only the "average" of the curvature is negative.

However, if we turn to higher-dimensional spaces, then we can construct a closed surface S_g with $g \geq 2$ by tearing up a pseudo-sphere into little pieces and then gluing them back together in a certain manner. In a higher-dimensional space, S_g is a surface of constant curvature. Higher-dimensional spaces allow us to create mental images of new figures.

The geometry of higher-dimensional spaces has developed rapidly since the beginning of the twentieth century, and many significant concepts and theories have been generalized. Among them are the definitions of the Euler characteristic and the curvature in higher dimension. The exploration of generalized theories in higher dimensions became an important theme of geometry. The Poincaré–Hopf theorem and the Gauss–Bonnet theorem were expanded to higher dimensions and used to develop some of the most profound theories of modern mathematics.

These generalized theories in higher dimensions have led to an understanding of the following concepts: moduli spaces of Riemann surfaces, Teichmüller spaces, and the superstring theory, all of which are considered to be at the cutting edge of modern mathematics. Through these theories, the generalized Poincaré–Hopf and the Gauss–Bonnet theorems in higher dimensions are connected to the latest fields of modern physics.

Chapter 2

The Story of Dimension

Introduction

The readers perhaps have not often (or ever!) heard the word "dimension" during their math classes. The concept of dimension might, therefore, at first seem foreign and unfamiliar. However, if we think of a line as one-dimensional, a plane as two-dimensional, and space as three-dimensional, then our readers will likely grasp the concept quite readily.

If you extend your arm horizontally, then the extended arm shows the direction of a one-dimensional line. If you raise your arm straight up in the extended position, then your arm moves in a two-dimensional plane. If you swing your arm in a circle over your head, then that motion occurs in three-dimensional space. Indeed, the authors are convinced that the concept of dimension is firmly imbedded in our consciousness and is an important facet of our daily lives.

Similarly, distinguishing dimensions into one-dimension, two-dimension, and three-dimension, may be derived from the observation that a line has a "horizontal" direction, a plane has "horizontal and vertical" directions, and space has "horizontal, vertical, and depth" directions. As you can see, when the number of dimensions increases, the number of independent directions increases in succession. Intuitively, we sense that no matter how many times a one-dimensional string is "bunched up," the resulting figure is still one-dimensional. We realize it cannot become a two-dimensional figure. We also realize that an intricately folded rectangular piece of paper is still two-dimensional. These perceptions indicate that what we intuitively understand with regard to the concept of dimension is, in fact, much more complicated than it would appear on the surface. To truely understand this concept we must undertake a much deeper and rigorous investigation.

Mathematicians first began to ask the question "What is dimension?" at the end of the nineteenth century. This question arose after the introduction of the theory of sets by Georg Cantor (1845–1918) and the discovery of a strange curve called the Peano curve. After these discoveries, mathematicians began to explore the question of dimension in earnest. The theory of dimension was introduced in the 1920s, and thereby the mathematical viewpoint of dimension was established. It might be helpful to begin our journey to understanding the concept of dimension from a historical perspective.

It is quite reasonable for anyone to imagine that there should be four-dimensional space, even if one might think of it as strictly hypothetical. However, the dimensions that we can actually perceive are limited to three dimensions. The reader might think that four-dimensional space is a subject of a science fiction story, not a topic discussed in mathematics, but this is not necessarily so. To understand this, one should consider the fact that even though the previously mentioned intricately folded sheet of paper is a two-dimensional surface, the shape of the folded paper is observed in three-dimensional space. Similarly, to observe a three-dimensional surface, four-dimensional or higher-dimensional space is generally needed.

A two-dimensional sphere of radius 1 is represented by the equation $x^2 + y^2 + z^2 = 1$, using the space coordinates x, y, and z. With this representation, we understand that the sphere is round. If we follow this analogy, a three-dimensional sphere of radius 1 will be given by the equation $x^2 + y^2 + z^2 + w^2 = 1$, using four-dimensional space coordinates x, y, z, and w. What does this figure look like in four-dimensional space? You now see that to understand three-dimensional surfaces we need to have four-dimensional space.

When we observe a three-dimensional surface in four-dimensional space, we find new geometric images which are far beyond our ordinary perception of figures. We will discuss this later in this chapter. In Chapter 1 we studied closed surfaces in three-dimensional space from a global viewpoint. Here we would like the reader to elevate the thought process to the next higher dimension (the fourth dimension). This topic has not yet been fully explored or understood and it is one of the main subjects of study in modern topology.

We can say that dimension is the most fundamental quantity to understand space. With the progress of physics, our understanding of the concept of space has deepened, and the concept of dimension has been evolving continuously, in the field of physics. Geometry and physics are the two viewpoints from which we observe space. These viewpoints have converged rapidly towards each other in the twentieth century. Lately, it seems as though they have been fused and become one. The development of physics and the evolution of the concept of space are discussed in the Appendix.

LECTURE 1

Learning to Appreciate Dimension

1.1. Seeing Is Not Believing

We believe the reader knows that a line is one-dimensional, a plane is two-dimensional, and space is three-dimensional. If you have studied the theory of relativity in physics, you might know that space-time is four-dimensional. One might also encounter a four-dimensional universe in science fiction.

However, the question still remains, "What is dimension?" One dimension is represented by a line on which real numbers are associated with points of the line. So we can say that a real number is a one-dimensional quantity. Points on a plane are identified by ordered pairs (x, y) of real numbers through the use of an xy-coordinate system. Therefore, we can say that an ordered pair (x, y) expresses a two-dimensional quantity. A point in space is identified by an ordered triple (x, y, z) of real numbers. Therefore, it is considered a three-dimensional quantity. Similarly, we can say that an ordered quadruplet (x, y, z, w) represents a four-dimensional quantity, even though we do not know how to draw a picture of four-dimensional space. An ordered n-tuple $(x_1, x_2, \ldots, x_n)$ represents an n-dimensional quantity.

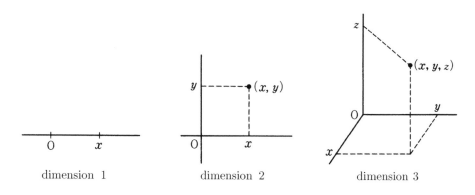

dimension 1   dimension 2   dimension 3

FIGURE 1.1

The set of all real numbers is denoted by $\mathbb{R}$. The set of all ordered pairs (x, y) of real numbers is denoted by $\mathbb{R}^2$. In general, the set of all n-tuples $(x_1, x_2, \ldots, x_n)$ of real numbers is denoted by $\mathbb{R}^n$. Around the middle of the nineteenth century the notion of dimension began to enter the minds of mathematicians. Mathematicians regarded n-tuples as multivariate quantities. The question of what properties define four-dimensional space never crossed their minds before. At that time, the object of geometry was to study relations of figures which could be visually observed. The differences between a line, a plane, and space are clear if you simply visualize them. Because of this visual observation, the notion of dimension seemed so intuitively obvious, that there was no reason to ask the question "What is dimension?"

Since the time of the Greeks, the concept of dimension was vaguely and inaccurately perceived as simply meaning that a plane contains more points than a line and space contains more points than a plane. It is only natural that the reader may also have the same vague and inaccurate idea of dimension.

This vague perception of dimension was shockingly disturbed by a significant discovery which occurred in 1877. Around this time, a German mathematician Georg Cantor had immersed himself in the idea of "counting infinite numbers."

"Counting infinite numbers" is explained as follows. The most fundamental concept of mathematics is that of natural numbers $1, 2, 3, 4, 5, \ldots$. Suppose you have five apples in one basket and an unknown number of oranges in another basket; you can determine the number of oranges by "pairing" one apple with one orange and placing them in a third basket (a one-to-one correspondence). If we finished "pairing" the apples and oranges and we have no oranges left, then we can say that we had five oranges in the basket. The number five was born out of one-to-one correspondences like this. Similarly, each natural number is obtained in the same manner. However, we do not deal with only finite objects in mathematics. Consider the set of all natural numbers

$$(1) \qquad\qquad \{1, 2, 3, 4, \ldots, n, \ldots\}.$$

It consists of infinitely many elements. The set of rational numbers also consists of infinitely many elements. The same is true for the set of real numbers. Therefore, if two sets with infinitely many elements are given, it is reasonable to think of counting infinite numbers by trying to establish a one-to-one correspondence of these two sets.

For example, the set of integers

$$(2) \qquad\qquad \{\ldots, -3, -2, -1, 0, 1, 2, 3, \ldots\}$$

seems to have more elements than the set of natural numbers (1). However, the two sets (1) and (2) are in one-to-one correspondence by the following rule:

$$
\begin{array}{ccccccccc}
1 & 2 & 3 & 4 & 5 & \ldots & 2n & 2n+1 & \ldots \\
\updownarrow & \updownarrow & \updownarrow & \updownarrow & \updownarrow & & \updownarrow & \updownarrow & \\
0 & 1 & -1 & 2 & -2 & \ldots & n & -n & \ldots
\end{array}
$$

Hence, the sets (1) and (2) have the same number of elements, that is, they both have a certain infinite number of elements. In mathematical terms, we say that the sets (1) and (2) have the same cardinal number (or cardinality). In general, if there exists a one-to-one correspondence between two sets, we say that these two sets **have the same cardinal number**.

Cantor originally used the symbol $\aleph_0$ to denote the cardinality of a set that has the same cardinal number as the set (1) of natural numbers. That is, any set which is in one-to-one correspondence with the set of natural numbers, has the **cardinal number** $\aleph_0$. The symbol $\aleph$ (pronounced aleph) is the first letter in the Hebrew alphabet. The symbol $\aleph_0$ is pronounced as aleph-null. Hence, we can say that the set of integers has the cardinal number $\aleph_0$.

CHALLENGE 1. Show that set of odd numbers has the cardinal number $\aleph_0$.

Cantor counted many different sets by using the one-to-one correspondence method. He showed that the set of rational numbers also has the cardinality $\aleph_0$. After this discovery, he spent a great deal of time thinking about the notion of infinity. He finally proved that the set of real numbers is not in one-to-one correspondence with the set of natural numbers. That is, the cardinality of the set of real numbers is not the same as the cardinality of the natural numbers. He showed that the cardinality of the set of real numbers is greater than the cardinality of the set of natural numbers. He decided to use the symbol $\aleph$ as the cardinal number of the set of real numbers. The set of real numbers is said to have the **power of continuum**.

♣ To show that the set of real numbers is not in one-to-one correspondence with the set of natural numbers, we will use the set of real numbers x such that $0 < x \leq 1$ instead of the set of all real numbers. So let our set be

$$S = \{x \mid 0 < x \leq 1\}.$$

We will show that S is not in one-to-one correspondence with the set of natural numbers by first assuming that S is in one-to-one correspondence with (1) and then showing that this assumption produces a contradiction. Now, suppose S is in one-to-one correspondence with the set of natural numbers:

$$
(3) \qquad
\begin{array}{ccccccc}
1 & 2 & 3 & \dots & n & \dots \\
\updownarrow & \updownarrow & \updownarrow & & \updownarrow & \\
x_1 & x_2 & x_3 & \dots & x_n & \dots
\end{array}
$$

where the set S is written as $\{x_1, x_2, x_3, \dots, x_n, \dots\}$. If we express each element of S as an infinite decimal expansion, we obtain the following:

$$x_1 = 0.\alpha_1\alpha_2\alpha_2\ldots\alpha_n\ldots$$
$$x_2 = 0.\beta_1\beta_2\beta_3\ldots\beta_n\ldots$$
$$x_3 = 0.\gamma_1\gamma_2\gamma_3\ldots\gamma_n\ldots$$
$$\cdots\cdots\cdots\cdots\cdots\cdots\cdots$$
$$x_n = 0.\lambda_1\lambda_2\lambda_3\ldots\lambda_n\ldots$$
$$\cdots\cdots\cdots\cdots\cdots\cdots\cdots$$

(A finite decimal number such as 0.3 is expressed by an infinite decimal number $0.2999\ldots$.)

If we now consider an infinite decimal number

$$\tilde{x} = 0.\omega_1\omega_2\omega_3\ldots\omega_n\ldots,$$

where ω_1 is a number between 1 and 9 different from α_1; ω_2 is a number between 1 and 9 different from β_2; $\ldots$; ω_n is a number between 1 and 9 different from λ_n; and so on, then $\tilde{x}$ is not equal to x_1 (since their first decimal digit is not the same), $\tilde{x}$ is not equal to x_2, $\ldots$, $\tilde{x}$ is not equal to x_n (since their nth decimal digit is not the same), and so on. Therefore, $\tilde{x}$ is a number in S that does not appear in (3). This is a contradiction. Thus it is shown by contradiction that S is not in one-to-one correspondence with the set of natural numbers.

This was an astonishing discovery. Until then, infinity was regarded as something which is not finite. It was understood as a negation of finiteness. For the first time in history, Cantor showed that the cardinalities of infinite sets are not always the same. He showed that there are different kinds of infinity. In fact, the infinity of cardinality $\aleph_0$ is not the same as the infinity of cardinality $\aleph$!

After showing that the set of real numbers is not in one-to-one correspondence with the set of natural numbers, Cantor decided to study the set of all points on a line and the set of all points on a plane. He probably expected that the cardinality of the set of all points on the plane would be greater than $\aleph$, and the cardinality of the set of all points in space even greater than that of a plane. Thus, he assumed that the cardinality of the set of points increases as we move from a line to a plane and then from a plane to space. He expected that the property of dimension might be understood by observing the cardinality of different sets. However, his expectation was utterly betrayed. In fact, the set of all points on the plane as well as the set of all points in space are in one-to-one correspondence with the set of all points on a line. He discovered that the sets of all points of a line, of a plane, and of space all have the same cardinality $\aleph$.

NOTE. This is the discovery that caused the "shocking disturbance" referred to previously. The vague concept of dimension in our minds, which had been with us since ancient times, had fallen apart.

In July of 1877 Cantor wrote a letter to Dedekind to inform him of this discovery. In this letter, he wrote that he had proved that the set of all points on a line is in one-to-one correspondence with the set of all points on a plane (actually, Cantor considered a more general case where the correspondence was between $\mathbb{R}$ and $\mathbb{R}^n$), and he asked Dedekind to determine whether there were any errors in his proof. Cantor expressed his astonishment with regard to this discovery thusly in French,

"*Je le vois, mais je ne le crois pas*" (I see it, but I do not believe it!).

♣ This discovery, which astonished Cantor, can be illustrated by the following example.

Consider a plane completely covered by sand without any open space. Suppose the sand is scooped up and put in a bag which has a hole in the bottom so small that only one grain of sand at a time can pass through the hole. If one walks along the number line with the bag on the shoulders, grains of sand would drop one by one through the hole and would eventually cover the number line. Thus, the sand on the plane is in one-to-one correspondence with the set of grains of sand on the plane. Cantor proved this fact by using the square $\{(x,y)\colon 0 < x \le 1, 0 < y \le 1\}$ on the plane as follows:

He expressed the coordinates of a point (x,y) in the square as $x = 0.\alpha_1\alpha_2\alpha_3\ldots$ and $y = 0.\beta_1\beta_2\beta_3\ldots$. To that point (x,y) he associated the point $0.\alpha_1\beta_1\alpha_2\beta_2\alpha_3\beta_3\ldots$ on the interval $(0,1]$ of the number line and continued the proof by contradiction, as before (it may be interesting to note that this proof requires minor revision).

In any case, this discovery caused a great stir in the world of mathematics. The vague concept of dimension was dispelled and along with it the ambivalence toward the actual meaning of dimension. Mathematicians began to ask, "What precisely is the dimension?"

1.2. Peano Curve

The fascinating fact that the set of points of a line is in one-to-one correspondence with the set of points of a plane depends on the assumption that points on the plane are considered disconnected from each other. However, when we think about a line or a plane, we think of them as a series of connected points. We see that points on a line are continuously connected and points on a plane are continuously connected in all directions. The continuous connection of points on a line in either direction seems to exhibit the characteristic property of a one-dimensional space, while the continuous connection of points on a plane in all directions seems to exhibit the characteristic property of a two-dimensional space. Therefore, if we consider the notion of dimension from the point of view of continuity, we cannot conclude that a line and a plane are the same object.

However, in 1890, the Italian mathematician Giuseppe Peano (1858–1932) showed that something beyond our common knowledge (or intuition) occurs when we look at a line and a plane. It was a discovery which caused a great reaction among the mathematicians of that time. Peano showed the following fact.

◘ There exists a continuous mapping φ from the closed interval $[0, 1]$ of the number line onto a unit square (a square with sides of unit length) such that the image of φ is the set of all points of the square.

♣ We need to explain some of the terms we introduced in the above statement. The closed interval $[0, 1]$ of the number line is the set of all real numbers x such that $0 \leq x \leq 1$. A mapping φ, which is a Greek letter pronounced "fi", is a mapping which associates with x in $[0, 1]$ a point P in a square and we write $P = \varphi(x)$. When we say that a mapping φ is continuous, we mean that it has the property that the sequence of points $P_n = \varphi(x_n)$ approaches the point $\varphi(x_0)$ whenever a sequence of points x_n $(n = 1, 2, \dots)$ on $[0, 1]$ approaches the point x_0.

The diagram in Figure 1.2 might help you understand the concept of continuous mapping.

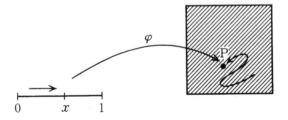

FIGURE 1.2

Peano's result is quite the opposite to what common sense tells us. For example, from our experience, we know it is impossible to entirely cover a square with an elastic string of length 1 cm no matter how we manipulate (stretch, twist, etc.) the piece of elastic. However, if we look at this problem from the mathematical point of view, Peano showed that it is possible to cover an entire square with a line (an elastic string). How can a phenomenon which is seemingly contradictory to common sense occur? The answer is simply that the phenomenon is outside the boundaries of our common sense, which is limited to finite manipulation (stretching and twisting, etc.) of a string. However, as we know, a real line contains an infinite number of points. Using the notion of the limit of a sequence of manipulations, it is possible to find a mapping φ from $[0,1]$ to the square such that the image of φ covers the entire square.

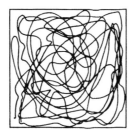

FIGURE 1.3

The mapping φ described earlier in ◪ is called a **Peano curve**. We now know that in fact there exist many Peano curves. We construct the well-known and widely used Peano curve which was originally constructed by Peano himself.

We explain the construction by an example using the diagrams in Figure 1.4. Let the variable x be time, varying in the closed interval $[0,1]$, which represents a fixed one hour time interval. Suppose a square represents a park which a policeman must patrol in one hour. The policeman first divides the park into four smaller blocks and patrols each of the blocks for 15 minutes ($= 1/4$ hour). The diagram of his patrol pattern is shown by the mapping φ_1 in Figure 1.4(I). The policeman starts at the point P_0 and reaches P_1 in fifteen minutes. Fifteen minutes later he reaches P_2, and fifteen minutes later, he reaches P_3. Finally, in one hour from the start, he reaches P_4 and completes his patrol of the park.

However, should the policeman need to patrol the park a little more extensively, he might further divide each of the four blocks into four smaller blocks (totalling 16 smaller blocks). In that situation, he would patrol four smaller blocks in fifteen minutes, and then proceed to the next four blocks spending fifteen minutes patrolling those blocks, and so on for the remaining blocks. If he patrols the park in such a manner, the patrolling pattern becomes a bit more complicated (as shown by the mapping φ_2 in Figure 1.4(II)). Now suppose that to tighten the security of the park patrol even further, each smaller square is divided yet again into four smaller blocks to patrol. There are $4^3 = 64$ smaller blocks in the park now. An efficient patrol pattern which does not alter the general direction of the original patrol pattern of Figure 1.4(I) is shown by the mapping φ_3 in Figure 1.4(III). The black dots in the diagram represent the locations at which the policeman arrives every $1/64$ hour. By subdividing the patrolling blocks into finer sub-blocks, the policeman can examine the area in greater detail (including the

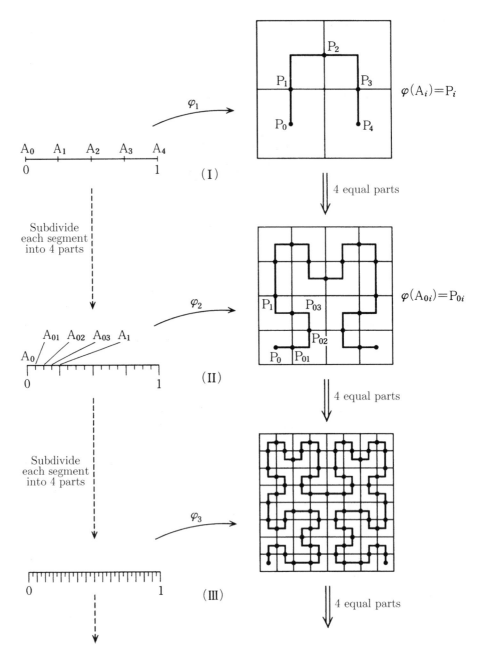

FIGURE 1.4

bushes). If a thief is hiding in a bush, he would probably see the policeman walking very near him! However, if the policeman checks a block every $1/4^n$ hour with a patrolling pattern of 4^n blocks determined by a mapping φ_n with a larger n, there would still be a place for a thief to hide (considering the thief to be represented by a point). What we can be sure about is that if we make n larger and larger, then the policeman can get closer and closer to the thief.

Therefore, if a mathematician looks for an ultimate "patrolling" pattern for the policeman, he will take the mapping $\varphi(x) = \lim_{n\to\infty} \varphi_n(x)$; then the patrolling path will pass through every point in the park, and there will be no place for the thief to hide (again, provided the thief is represented by a point).

The map $\varphi(x)$ obtained in this manner is, mathematically, a continuous mapping from the closed interval $[0,1]$ to the unit square, and for every point P in the square, there exists x ($0 \le x < 1$) such that $\varphi(x) = P$. This mapping φ is the Peano curve we spoke about previously.

The explanation here is not totally exhaustive, but if you get the general idea of what a Peano curve is, that is enough. The important point here is to understand that the thief might see the policemen walking near him more than once (depending on where he is hiding) if the patrolling pattern is as shown in Figure 1.4(III). This means that as n approaches infinity, $\varphi(x) = \varphi(x')$ for two different time moments x and x'. In fact, on a Peano curve, you will find double, triple, or quadruple points. A quadruple point p is a point in the square such that

$$P = \varphi(x_1) = \varphi(x_2) = \varphi(x_3) = \varphi(x_4)$$

for four different time moments x_1, x_2, x_3, and x_4. It is essential to recognize that there are multiple points in the square. In fact, the set of multiple points is dense in the unit square.

LECTURE 2

What is Dimension?

2.1. Poincaré's Idea

We have learned from the existence of Peano curves that a line segment cannot be distinguished from a square merely by observing continuously connected points. However, even when we acknowledge Peano's finding, our intuitive geometric concept of dimension tends to remain unshaken. There must be an idea which supports the concept of dimension, and thereby bridges the gap between Peano's finding and our intuition. Such an idea is that of a one-to-one continuous mapping.

A **one-to-one continuous mapping** φ is a continuous mapping such that if $x \neq x'$ then $\varphi(x) \neq \varphi(x')$, and if $x_n \to x$ then $\varphi(x_n) \to \varphi(x)$ $(n \to \infty)$. Peano curves lack this one-to-one property.

Under a one-to-one continuous mapping, a line segment cannot be sent to a square, nor a square to a cube. If we restrict our discussions of mapping to a two-dimensional plane, then a one-to-one continuous map may send a line segment to a complicated curve. However, this curve cannot completely cover a region on the plane (for example, the interior of a small circle). To our eyes, the curve still appears to be a one-dimensional figure. If we consider this kind of mapping in the case of space, then a one-to-one continuous mapping may send a square to a surface such as the one obtained by puncturing the complicated closed surface with Euler characteristic 0 (as discussed in Chapter 1). However, this surface cannot completely occupy a region in space (for example, the interior of a sphere). A surface is still a two-dimensional figure.

Next, we should explore what one-dimensional, two-dimensional, three-dimensional, and n-dimensional figures are.

The first person who addressed the question "What is Dimension?" was the great French mathematician Henri Poincaré (1854–1912). In 1905, in the publication "The Value of Science," Poincaré discussed his ideas regarding dimension. We will briefly summarize the main points Poincaré stated in this publication. First, he made the notion of "cuts" to be his central theme. Poincaré stated that a continuum of one dimension can be divided into two pieces by a point. A continuum of two dimensions is the one that can be divided into two pieces by a one-dimensional line or a curve. A continuum of three dimensions is the one that can be divided into two regions (the inside and the outside) by a two-dimensional surface.

For clarity, we illustrate the above concept as follows. First, Poincaré stated that a characteristic property of a one-dimensional figure is that it can be divided into two pieces by a point. This means a tangled string is one-dimensional because it can be divided into pieces by cutting the string at several points. Suppose a man is living in a one-dimensional world. The world he lives in is represented by a line. Anything other than the line he lives in is outside of his universe. Now, suppose he is a rich man and he has a treasure trove. The only place he can hide his treasure is also on this line. If he wants to protect his treasure, he must construct barriers (possibly even double or triple barriers as in Figure 2.1). Since he lives in a one-dimensional world, the barriers are points as in Figure 2.1. Intruders can be shut out by these point "barriers." This is typical for a one-dimensional world.

Point doors   Treasure   Point doors   Circular fences   Treasure

Treasure

FIGURE 2.1

Next, consider a rectangular piece of paper having four sides. If we cut out a figure using a pair of scissors, then the resulting figure will be two-dimensional. It is two-dimensional because such a figure can be further cut into smaller pieces. Each time the paper is cut, the cut goes along the trace on the paper, where the scissors are applied. This cut is a curve, which is a one-dimensional figure. Imagine a rich man living in this two-dimensional plane, who owns a great treasure. If he wants to protect this treasure, it would be a good idea to draw several circles around it. These circles will protect it. This is typical for a two-dimensional plane (or world).

A solid body can be divided into several regions by slicing it with several surfaces. Since each cross-section is a two-dimensional surface, this means that the body is three-dimensional. We live in such a three-dimensional world. A rich man living in a three-dimensional world would hide his treasure inside a safe (a box) surrounded by several layers of protective surfaces. A surface divides a solid body into two regions (interior and exterior). This is typical for a three-dimensional world.

Poincaré stated that the four-dimensional world is the one in which a three-dimensional solid body would divide a figure into two regions (the inside and the outside). A rich man living in a one-dimensional world could never protect his treasure with barriers (represented by points) from people who live in a two-dimensional world. The people living in a two-dimensional world could see where the treasure was hidden by looking at it from above or below or from any angle not on the one-dimensional line and could reach the treasure from these angles. Furthermore, a rich man living in a two-dimensional world could never protect his treasure with barriers (represented by circles) from people who live in a three-dimensional space: they could see where the treasure was hidden by looking at it from either above or below the two-dimensional plane. If we extend this idea to a four-dimensional world, it might be possible that the people living in a four-dimensional world could look inside the rich man's safe from a direction other than horizontal, vertical, or depth. This sounds like a science fiction story, doesn't it?

Poincaré explored this idea further, and in the last year of his life (1912), he published a book titled "Dernières pensées" (Last Essays). In Chapter 3, Section 2 of this work, he clarified his idea of dimension. Section 2 of his book begins with the question "What is a continuum of n dimensions?" The analytic definition of a continuum of n dimensions is given as a set of all n-tuples which satisfy certain analytic inequalities. However, Poincaré made two critical observations about using n-tuples to define a continuum of n dimensions. The first observation was that the coordinates in the space are not fixed, but rather can be changed freely. Therefore, he believed that the definition of a continuum of n dimensions should be approached from a more basic point of view than the use of n-tuples. The second observation was that the conventional n-tuple definition underestimates the intuitive origin and rich and abundant content of the notion of continuity. Poincaré's criticism of using n-tuples to define a continuum of n dimensions was very severe indeed! In his book, Poincaré pointed out that we should pay close attention to inherent mathematical intuition when we encounter formal definitions of concepts. His ideas provided an important impetus to the development of the theory of dimension.

Poincaré explained how to construct a definition of dimension as follows.[1] "I shall base the determination of the number of dimensions on the notion of *cuts*. Let us consider first of all a closed curve, that is, a continuum in *one* dimension. If on this curve we take any two points through which we shall not permit ourselves to pass, the curve will be cut into two parts, and it will become impossible to go from one to another still remaining on the curve but not passing through excluded points. Let us consider, on the other hand, a closed surface, which forms a continuum of *two* dimensions. It will be possible to take on this surface one, two, or any number of excluded points whatever. The surface will not be divided into two parts because of this; it will be possible to go from one point to another on this surface without encountering any obstacle because it will always be possible to *go around* the excluded points.

But if we trace on the surface one or many closed lines and if we consider them as *cuts* which may not be crossed, the surface can then be cut into several parts."

Similarly in the case of three-dimensional space, two-dimensional cross-section cuts are needed to divide the space into regions. As an extension

[1]H. Poincaré, *Mathematics and Science: Last Essays. Dernières pensées*, Dover Publ., New York, 1963.

of the above thought, Poincaré defined a general n-dimensional space as follows.

"A continuum has n dimensions when it is possible to divide it into many regions by means of one or more cuts which are themselves continua of $n-1$ dimensions. The continuum of n dimensions is thus defined by the continuum of $n-1$ dimensions. This is a definition by recurrence."

Based on Poincaré's philosophy of dimension and the newly developed abstract topology, the general theory of dimension was established in the 1920s. Hopefully, at this point our readers have come to a general understanding of the concept of dimension.

♣ A topological space is a set with a concept of nearness. A general definition of a topological space is rather abstract to describe here, so we will instead give you a definition of a metric space. Let M be a set and d a real-valued function on $M \times M$ such that for every x, y, and z in M, the following conditions are satisfied:

(i) $d(x,y) \geq 0$ and $d(x,x) = 0$;

(ii) $d(x,y) = d(y,x)$;

(iii) $d(x,y) \leq d(x,z) + d(z,y)$. Then the function d is called a metric on M, and the set M with the function d is called a metric space. A sequence $\{x_n\}$, $(n = 1,\ 2,\ \dots)$ of points in a metric space converges to a point x if $d(x_n,x)$ has limit 0 as n increases without bound $(d(x_n,x) \to 0$ as $n \to \infty)$. Hence, in a metric space the concept of limit can be introduced.

For any point x in a metric space M the set of all points whose distance from x is less than $\frac{1}{2}$ is called the $\frac{1}{2}$-neighborhood of x and is denoted by $V_{\frac{1}{2}}(x)$. In general, for any positive number ε, the ε-neighborhood of x, $V_\varepsilon(x)$, is defined by

$$V_\varepsilon(x) = \{y \mid d(x,y) < \varepsilon\},$$

and the boundary of $V_\varepsilon(x)$, denoted by $\operatorname{Bdry} V_\varepsilon(x)$, is defined by

$$\operatorname{Bdry} V_\varepsilon(x) = \{y \mid d(x,y) = \varepsilon\}.$$

The dimension of a metric space M denoted by $\dim M$ is defined inductively as follows.

(i) A metric space M that consists of only a point or several discrete points (such that the distance between any two points is 1) by definition has dimension 0. This is denoted as follows: $\dim M = 0$.

(ii) Suppose $(n-1)$-dimensional spaces are defined. If for any x in M and any small positive number there exists an even smaller positive number ε such that

$$\dim \operatorname{Bdry} V_\varepsilon(x) \leq n - 1,$$

then we say that $\dim M \leq n$.

(iii) If $\dim M \leq n$ holds, but $\dim M \leq n-1$ does not hold, then we define the dimension of M as n, which is denoted by $\dim M = n$.

According to this definition, one-dimensional space, two-dimensional space, and higher-dimensional spaces are defined in succession. If there is a one-to-one continuous mapping from a metric space M to another metric space N, then a point of M is mapped to a point of N, a neighborhood of that point of M is mapped to a neighborhood of its image in N, and the boundary of the neighborhood in M is mapped to the boundary of the neighborhood in N. Thus, the dimension does not change (is invariant) under such a mapping.

2.2. Lines, Planes, and Spaces

We think of a line as being one-dimensional, a plane as being two-dimensional, and space as being three-dimensional. The difference in dimension naturally gives different characteristics to a line, a plane, and space. Let us discuss some topics related to these characteristics.

I. Flagstone Theorem. [2] Let us introduce a new concept, that of an open set. A subset V of a one-dimensional line is called an open set if for any point x in V, "every point in the immediate vicinity of x is again in V." A subset V of a two-dimensional plane is called an open set if for any point x in V, "every point in the immediate vicinity of x is again in V." Therefore, an open set V on a plane is a set such that for any point x in V, there exists a very small disk centered at x which is completely contained in V. An open set on a curve and that on a surface are similarly defined (Figure 2.2).

x

An open set V on a
one-dimensional line

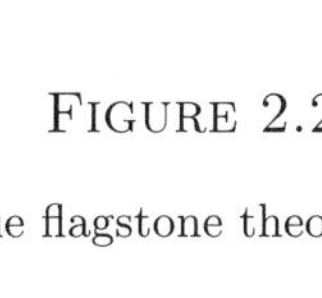

An open set V on a
two-dimensional plane
(the boundary is not included)

FIGURE 2.2

---

[2]It is sometimes called the Lebesgue flagstone theorem.

Now, let us imagine laying flagstones made from open sets on a line (or a curve), a plane (or a surface), or space. There will be places where the open sets overlap. We want to focus our attention on the number of layers of these overlaps with the following two conditions in mind:

(a) The flagstones can be extremely small.
(b) The smallest possible number of flagstones should be used to cover each point.

The basic question posed by the flagstone theorem is illustrated by the following question. If, according to condition (b), we try to minimize the number of flagstones used to cover each point, that is, to minimize the number of overlapping layers, what is the minimum number of overlapping layers of flagstones?

The answer to this question is as follows.

At some points of a one-dimensional line, two falgstones necessarily overlap. On a two-dimensional plane, three flagstones necessarily overlap. In a three-dimensional space, four flagstones overlap. If we extend this theorem to the general dimension n, then at least $n+1$ flagstones necessarily overlap in an n-dimensional space. This is known as the flagstone theorem. Let us then examine the conclusions of this theorem.

On a one-dimensional line we cannot lay flagstones without overlapping while still satisfying both conditions (a) and (b). This is shown by illustration (A_1) in Figure 2.3. If it were possible to steadily reduce the size of the flagstones and lay them without allowing overlaps, then the end result would be a set of discrete points rather than a continuous line. However, since our set is a continuous line, one would need to allow overlaps when laying flagstones. According to illustration (A_2) in Figure 2.3, we need at least two overlapping layers at some points on the one-dimensional line. We can see that any point on the one-dimensional line is covered by either one flagstone or two flagstones. Therefore the conclusion of this theorem for a one-dimensional line is that two flagstones must overlap. On a two-dimensional plane we cannot lay flagstones without allowing an overlap of more than two stones while still satisfying both conditions (a) and (b). This can be verified by examining illustration (B_1) in Figure 2.3. If we continue to reduce the size of flagstones, only a line can be covered by them. However, if we allow overlaps of three flagstones, then we can cover the entire plane. This can be verified by examining illustration (B_2) in Figure 2.3. Hence the conclusion of the theorem for a two-dimensional plane is three flagstones.

The pattern of placing flagstones on a plane shown in Figure 2.3(B_2) is rather unique. It is not obvious, at a glance, how to extend this pattern to a three-dimensional space. However, there exists another pattern of placing flagstones which is related to the concept of triangulation discussed in Lecture 1 of Chapter 1. This pattern allows overlaps of three stones. These arguments become clear as we investigate this pattern further.

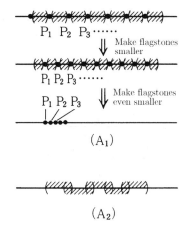

(A_1)

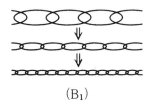

(A_2)

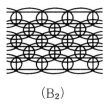

(B_1)

(B_2)

FIGURE 2.3

We know that a plane or a surface can be triangulated. Consider a surface first. It is possible to further subdivide a triangulated surface using the barycentric subdivision. In fact, using this method we can make the subdivisions as small as desired. A triangle in this triangulation is not an open set since it contains the boundary (sides and vertices); therefore, it cannot be used as a flagstone. However, if we consider a triangle ABC in the triangulation of a surface and the union of all triangles that have A as a vertex, then the interior of this figure is an open set. This set is called the star neighborhood of the vertex A and is denoted by $\text{st}(A)$. Star neighborhoods $\text{st}(B)$ and $\text{st}(C)$ of the vertices B and C are defined similarly. Taking star neighborhoods for each triangle in the triangulation, we can regard them as flagstones, and they will cover the surface. In this pattern, whenever there is an overlap, it involves at most three flagstones.

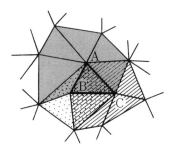

FIGURE 2.4. st(A) is the shaded region; st(B) is the dotted region; and st(C) is the lined region.

CHALLENGE 2. Show that st(A), st(B), and st(C) overlap if A, B, and C are the vertices of a triangle in a triangulation of a surface, and the overlapping region

$$\mathrm{st}(A) \cap \mathrm{st}(B) \cap \mathrm{st}(C)$$

is the set of all interior points of the triangle ABC.

The number 3 which appears in a two-dimensional example of the flagstone theorem, is closely related to the syllable "tri" in the word "trianguia-tion".

A space can be divided into tetrahedra instead of triangles. If we consider a triangulation of a space by tetrahedra and cover the space with star neighborhoods of vertices of each tetrahedron, we can see that at most four star neighborhoods overlap. With the above explanation, the connection between the dimension and geometry, which is the basis of the flagstone theorem, should become clearer.

CHALLENGE 3. Consider the questions discussed in Challenge 2 from the point of view of triangulation of a space rather than a surface (plane).

The inductive definition of dimension suggested by Poincaré has one problematic point. Namely, the dimension n cannot be directly related to the intrinsic geometric properties of the n-dimensional space. Because of this problem with Poincaré's definition, the flagstone theorem is sometimes used to define the dimension. We let the dimension of a space be n if $n + 1$ is the number satisfying the flagstone theorem. It may be rather difficult to grasp the intuitive concept of dimension with this definition, and in that sense this definition is intended for those who specialize in advanced concepts. Whichever definition of dimension is used, the end result is identical if the object we consider is an ordinary object.

II. Four Color Problem. The four color problem is a well-known problem involving coloring a map. This problem arose from the following practical question: What is the minimum number of colors needed to color the countries on a map on a plane or on a sphere so that no two countries sharing a common border will have the same color? (we consider all countries' areas to be connected and contiguous). If several countries intersect at a point as in Figure 2.5, we can ignore this point and color the remainder of the map.

FIGURE 2.5

CHALLENGE 4. Color the map in Figure 2.5 using the smallest number of colors so that adjacent countries have different colors. Next, create a more complicated map, and color it using the smallest number of colors.

You should find that the map in Figure 2.5 and more complicated maps you created can be colored by using four colors (if you have used more than four colors to paint the map, try again). No matter what kind of map you create, it becomes evident by experimenting that four different colors are sufficient to color the map so that no two adjacent countries will have the same color. These experiments lead to the following mathematical question: Is a set of four colors sufficient to paint a complicated map which has millions or even billions of countries on it? This question first arose in the middle of the nineteenth century, and in 1878 was formally posed by the British mathematician Arthur Cayley (1821–1895). This problem became known as the four color problem. Many mathematicians since then tried to solve it. The four color problem is easy to understand, and anyone can experiment with it by coloring a map; thus, it elicited a great deal of interest among a large number of people. However, it soon became quite evident, even to mathematicians, that it is, in fact, a very difficult problem to solve! It was possible to prove that a set of five colors is sufficient to color a map, by using induction on the number of countries on the map. However, when the question was asked on whether it is possible to delete one color from the set of five colors, no one could provide an answer. In fact, there was not even a hint of a proof in sight. It was conjectured at that time that there might have existed a map which would require five colors.

However, the four color problem was solved between the end of the 1970s and the beginning of the 1980s. It was proved that it is possible to color any map, no matter how complicated, with a set of four colors, as long as the map is on a plane or a sphere. At the end of the last stage of the proof of this problem, Kenneth Apple and Wolfgang Haken checked (by employing a supercomputer over an extended period of time) an enormous number of maps and verified that it is possible to color any of them with four colors. This is the only known proof of the four color theorem to date. Without the development of computers, it is possible that the four color problem would remain unanswered even today.

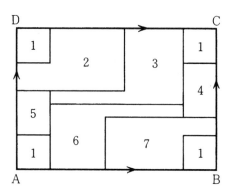

FIGURE 2.6

♣ A similar problem occurs when coloring a map on a torus. It is known that seven colors is needed to color a map on a torus. According to the Mathematical Dictionary published by Iwanami Book Publishing Company, the complete result on the map coloring on a surface of genus $g \geq 1$ was obtained by 1970. We know that six colors are necessary and sufficient to color a map on a Klein bottle. For surfaces of genus $g \geq 1$ other than the Klein bottle, the number of colors needed is given by the equation

$$\left[7 + \frac{\sqrt{49 - 24\chi}}{2} \right].$$

This number does not depend on orientability of the surface. In this expression, the Euler characteristic χ is given by

$$x = \begin{cases} 2 - 2g & \text{(for an orientable surface)}, \\ 2 - g & \text{(for a nonorientable surface)}. \end{cases}$$

The symbol [ ] used in the above expression is called the Gauss symbol, and it represents the greatest integer which is less than or equal to the number in the brackets.

The reason the four color problem was introduced in this chapter on dimension, is that this problem is typical for a two-dimensional plane.

To color a map on a one-dimensional line, only two colors are necessary and sufficient. This fact becomes obvious by looking at Figure 2.7.

FIGURE 2.7. Coloring countries on the line.

However, when we consider coloring a three-dimensional map in space, there exists a map that requires an infinite number of colors. No matter how many colors are prepared, it would not be possible to completely paint the three-dimensional map shown in Figure 2.8.

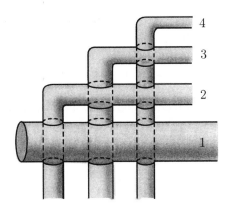

FIGURE 2.8. Coloring countries in three-dimensional space; 1, 2, 3, and 4 represent three-dimensional countries on a three-dimensional map.

As you have seen, if we look at the familiar figures of a line, a plane, and a space, it is the dimension that reveals the unique characteristic properties of each figure.

III. Fixed Point Theorem. A line and a plane look differently. Suppose a theorem is true for both a line and a plane. Even though the theorem might be quite easy to prove in the case of a line, it is often much more difficult to prove it in the case of a plane. Dimension, therefore, becomes an obstacle to the proof. We will now introduce Brouwer's fixed point theorem as a typical example of such a situation.

First, let us define a closed disk. The closed disk (that includes the boundary) of radius 1 centered at the origin on the coordinate plane is defined as the set of all points (x, y) on the plane that satisfy the following inequality:

$$x^2 + y^2 \leq 1.$$

If we use the phrase "the closed disk of radius 1 centered at the origin on the line," this means the set of all points x on the line satisfying the following inequality:

$$x^2 \leq 1,$$

that is, the set of all points such that

$$-1 \leq x \leq 1.$$

The following theorem is called Brouwer's fixed point theorem.

BROUWER'S FIXED POINT THEOREM. *Let B be the closed disk of radius 1 and φ a continuous mapping from B to B. Then there exists a point P_0 in B such that*

$$\varphi(P_0) = P_0.$$

In essence this theorem states the following:

> Suppose there is a circular tray whose surface is completely covered by sand. If the tray is shaken slightly (but continuously), each grain of sand on the tray also moves continuously to a new location. Naturally, some grains may overlap. If we represent each grain of sand as a point P of the closed disk, then we can consider the new location of the point P to be $\varphi(P)$. The fixed point theorem guarantees that no matter how we shake the tray, there exists at least one grain of sand which does not move from its original location.

As you can see from the above example, what the fixed point theorem states is not intuitively obvious. L. E. J. Brouwer (1881–1966) provided the proof of the fixed point theorem in 1910. His work gave birth to algebraic topology, a field that has developed rapidly since that time. Even today, to prove this theorem we must utilize the notion of homology, and it takes a substantial effort to do so.

We will explore the proof of this theorem in the case of one dimension. The fixed point theorem for one dimension is stated below (referring to the one-dimensional closed disk as discussed earlier).

If φ is a continuous mapping of the interval $I = [-1, 1]$ to itself, then there exists at least one point x_0 in I such that:

$$\varphi(x_0) = x_0.$$

In the case of one dimension, unlike that of two dimensions, we can immediately see how the theorem is proved. For this, we use the notion of the graph of a function, as in Figure 2.9. Since φ is a continuous mapping from I to I, it is a continuous function φ defined on the interval $I = [-1, 1]$ and taking values between -1 and 1, inclusive. Let us look at the graph of this function. From Figure 2.9 it is clear that the graph of φ intersects the diagonal line of the square with side I at a certain point P. Let the

x-coordinate of the point P be x_0. Then $\varphi(x_0) = x_0$, and the point x_0 is a fixed point of φ.

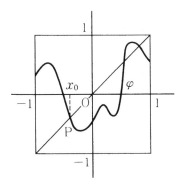

FIGURE 2.9

♣ To give a precise proof of the theorem, we need to use a theorem called the intermediate value theorem for continuous functions. Applying the intermediate value theorem to the continuous function $\Phi(x) = \varphi(x) - x$, we see that there exists a point x_0 in the interval I such that $\Phi(x_0) = 0$, that is, $\varphi(x_0) = x_0$.

Why is it so difficult to prove this theorem in the two-dimensional case, while it is quite easy to prove for the case of one dimension? The reason why an analog of the proof used for the case of one dimension cannot be applied to the case of two dimensions is that the graph of a continuous mapping from the closed disk to itself cannot be drawn. To be able to draw the graph of such a mapping, four-dimensional space is needed. Perhaps a person living in a four-dimensional space could visualize the reason behind the fixed point theorem for the case of two dimensions—it is not for us, simple mortals, to know!

♣ The fixed point theorem remains valid in dimension three. In fact, it is valid for any dimension n. The theorem in the case of n dimensions is given as follows:
Let B be the unit ball in $\mathbb{R}^n$ defined by

$$B = \{(x_1, x_2, \ldots, x_n) \mid x_1^2 + x_2^2 + \cdots + x_n^2 \leq 1\},$$

and let φ be a continuous mapping from B into B. Then there exists a point P_0 in B such that

$$\varphi(P_0) = P_0.$$

There are numerous applications of this theorem, and it is considered to be one of the fundamental theorems in modern mathematics. This theorem is used to show the existence of solutions of differential equations. Another application of it is in mathematical economics, where it is known as part of theory of equilibrium.

IV. Lengths of Sides and Diagonals. When we compare the coordinates of a square D_2 and a cube D_3 in space,

$$D_2 = \{(x_1, x_2) \mid -1 \leq x_1 \leq 1,\ -1 \leq x_2 \leq 1\}$$

and

$$D_3 = \{(x_1, x_2, x_3) \mid -1 \leq x_1 \leq 1,\ -1 \leq x_2 \leq 1,\ -1 \leq x_3 \leq 1\},$$

respectively, the fundamental dimensional difference is not immediately apparent. However, if you calculate the length of a diagonal line from any one of the vertices to the origin, you obtain

$$\sqrt{2} = 1.41421\ldots \quad \text{for } D_2, \text{ and}$$

$$\sqrt{3} = 1.73025\ldots \quad \text{for } D_3.$$

This clearly illustrates that the "depth" of the cube is greater than that of the square (see Figure 2.10).

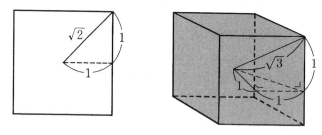

FIGURE 2.10

The higher the dimension, the greater the "depth". Consider an n-dimensional cube D_n in $\mathbb{R}^n$ defined by

$$D_n = \{(x_1, x_2, \ldots, x_n) \mid -1 \leq x_i \leq 1 \quad (i = 1, 2, \ldots, n)\}.$$

A half-diagonal of D_n, that is, a diagonal taken from any one vertex to the origin, has length

$$\sqrt{n}.$$

If the dimension of the space is 100, then the half-diagonal length of D_{100} becomes 10. This shows that the larger the number n, the greater the "depth" of the n-dimensional cube D_n. In a way, D_n is like a small closed box with each side of length 2. However, if the dimension n increases, so too does the depth, becoming even larger than the distance to the most distant points of the universe (see Challenge 5).

Now you may wonder why does the depth increase as the dimension gets higher? This phenomenon is apparently the "magic" occurring in high dimensions. Perhaps better put, this phenomenon clarifies the limitations of our ability to perceive higher dimensions due to the fact that we live and observe the space in three dimensions.

For a bit of fun, try the following challenge.

CHALLENGE 5. 1) Consider the square D_2 mentioned above, and imagine placing identical disks of maximum possible radius in each of the four quadrants of D_2. Then imagine placing a disk of maximum radius r_2 in the open space left around the origin (see Figure 2.11). Show that the radius r_2 of this circular disk around the origin is given by the formula

$$r_2 = \frac{\sqrt{2} - 1}{2}.$$

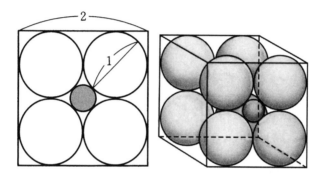

FIGURE 2.11

2) Consider the cube D_3 mentioned above. Imagine placing eight identical solid spheres of maximum radius to fill the cube in a manner similar to that in the previous problem. Imagine placing a solid sphere of maximum radius r_3 in the open space left around the origin. Show that the radius r_3 of this solid sphere around the origin is given by

$$r_3 = \frac{\sqrt{3} - 1}{2}.$$

3) Consider the ten-dimensional cube D_{10}. Imagine placing 2^{10} identical ten-dimensional solid spheres of maximum radius to fill the cube in a manner similar to that in the previous problem. Imagine again placing a ten-dimensional solid sphere of maximum radius r_{10} in the open space left around the origin. The radius r_{10} of this solid sphere is given by

$$r_{10} = \frac{\sqrt{10} - 1}{2} \approx 1.08113\ldots.$$

Show that this sphere around the origin is no longer contained within the cube, but will now extend beyond the boundary of the original ten-dimensional cube.

4) Consider the n-dimensional analogue of the above three problems. For an n-dimensional cube D_n, show that as $n \to \infty$, the n-dimensional solid sphere occupying the open space left around the origin will increase in size, eventually to the point of occupying the entire space $\mathbb{R}^n$.

2.3. Surfaces in Four-Dimensional Space

All the surfaces mentioned in Chapter 1 are two-dimensional surfaces, and we constructed them in the three-dimensional space $\mathbb{R}^3$. Suppose we constructed these surfaces in the four-dimensional space $\mathbb{R}^4$, where

$$\mathbb{R}^4 = \{(x, y, z, t) \mid x, y, z, t \text{ are real numbers}\}.$$

What would happen? To represent a point in $\mathbb{R}^4$, we need an additional variable t to be added to the $\mathbb{R}^3$ coordinates (x, y, z). Now, what does it mean to add this new coordinate t? To understand this, consider first the plane $\mathbb{R}^2$, where

$$\mathbb{R}^2 = \{(x, y) \mid x, y \text{ are real numbers}\}.$$

Then add another variable t (the coordinate representing the depth). We obtain the three-dimensional space $\mathbb{R}^3$, where

$$\mathbb{R}^3 = \{(x, y, t) \mid x, y, t \text{ are real numbers}\}.$$

Now let us examine in $\mathbb{R}^3$ the behavior of a curve drawn in $\mathbb{R}^2$.

Consider a curve C (see Figure 2.12) that crosses itself by making loops. Let P and Q be the points of self-intersection. Let us try to eliminate these points of self-intersection. One might try, for example, to pull a loop trying to "untie" it, or pull the strings trying to straighten the curve. However, such efforts would be futile because in the two-dimensional space, the intersection points cannot be undone regardless of how one manipulates the curve. Even if one were to, say, pull both ends of the "string" forming the curve such that the former loop(s) become cusp points, there still could be no straightening out (without destructing the curve) of the loop(s). The closest one could come to "pulling out" the loops would be to create those cusp points.

In $\mathbb{R}^2$

In $\mathbb{R}^3$

FIGURE 2.12

However, if we try to manipulate the curve in three-dimensional space, it is possible to eliminate the points of self-intersection by simply lifting the "upper string" at these points, that is, lifting part of the curve away from the points at which it crosses itself. This is possible since in the three-dimensional space, one can separate the upper and lower "string" because of the existence of altitude/depth. In other words, a self-intersecting curve in (x, y)-coordinate plane $\mathbb{R}^2$ can be changed into a non-self-intersecting curve in the (x, y, t)-coordinate space $\mathbb{R}^3$ by slightly moving one of the intersecting parts of the curve away in the positive direction of t, where t represents the altitude (depth) of a point in $\mathbb{R}^3$.

Recall that in Chapter 1, we mentioned that the Klein bottle is the surface which is constructed by identifying the opposite sides of a rectangle $ABCD$ according to the direction of the arrows (see Figure 2.13). The Klein bottle is a self-intersecting surface in $\mathbb{R}^3$ with no distinction between the interior and exterior sides. If we consider a Klein bottle in the four-dimensional space $\mathbb{R}^4$ by adding an extra variable t to (x, y, z), we can make the Klein bottle non-self-intersecting by simply moving one of the intersecting portions of the surface away in the positive direction of t. In other words, by passing from $\mathbb{R}^3$ to $\mathbb{R}^4$, we obtain the freedom of moving in space and we can undo the self-intersection.

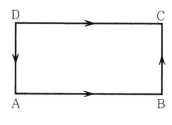

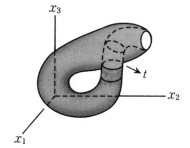

FIGURE 2.13. The Klein bottle.

The projective plane discussed in Chapter 1, can also be constructed as a non-self-intersecting surface in $\mathbb{R}^4$. To do so, we first create the projective plane as a self-intersecting surface in $\mathbb{R}^3$ called *Boy's surface*. Next, imagine placing it in $\mathbb{R}^4$. As you can see, we can undo this intersection in the same manner as we did for the Klein bottle in the previous figure. If you would like to examine Boy's surface in greater depth, see Chapter 6 of *Intuitive Geometry* by Hilbert and Cohn-Vossen. Additionally, construction of the projective plane in $\mathbb{R}^4$ is explained on pages 62–63 of *Differential Geometry* by Atsuo Sasaki.

♣ As we can construct a two-dimensional surface by pasting triangles, so can we construct a "three-dimensional surface" by pasting tetrahedra in such a way that every face of each tetrahedron is glued to a face of another tetrahedron. We call this "three-dimensional surface" a **three-dimensional manifold**. We can represent any three-dimensional manifold as a non-self-intersecting "three-dimensional surface" in $\mathbb{R}^6$. Can you imagine an $(n+1)$-hedron in the n-dimensional space $\mathbb{R}^n$? The standard $(n+1)$-hedron in $\mathbb{R}^n$ is a convex body in $\mathbb{R}^n$ with $(n+1)$ vertices at $(0, 0, \ldots, 0)$, $(1, 0, \ldots, 0)$, $(0, 1, 0, \ldots, 0)$, $\ldots$, $(0, 0, \ldots, 0, 1)$. By gluing these $(n+1)$-hedra, we can obtain an n-dimensional manifold. We can construct an n-dimensional manifold as a non-self-intersecting manifold by placing it in the 2^n-dimensional space $\mathbb{R}^n$.

Three-Dimensional Figures

3.1. Three-Dimensional Spheres

A sphere is the simplest two-dimensional surface, and it is a surface one can easily visualize. The standard sphere S^2 is the set of all points in $\mathbb{R}^3$ that satisfy the following equation:

$$(1) \qquad x_1^2 + x_2^2 + x_3^2 = 1.$$

What does the three-dimensional sphere S^3 look like? It is given by the set of all points in $\mathbb{R}^4$ that satisfy the following equation:

$$(2) \qquad x_1^2 + x_2^2 + x_3^2 + x_4^2 = 1.$$

We tend to imagine that a three-dimensional sphere exists as a type of round figure in the four-dimensional space $\mathbb{R}^4$. But our imagination may mislead us. It is impossible to make conclusions about the shape of a figure in the four-dimensional space using the analogy with shapes we see in our three-dimensional space. Instead, we must look at mathematical equations (1) and (2) in order to grasp what a three-dimensional sphere is like.

First, let us look at equation (1). In the two-dimensional sphere S^2,

the set of points that satisfy (1) and have $x_3 \geq 0$, represents the northern hemisphere of S^2,

and

the set of points that satisfy (1) and have $x_3 \leq 0$, represents the southern hemisphere of S^2.

These two hemispheres are glued together along the "equator" where $x_3 = 0$ (that is, they are glued together along the circle which is the set of all points that satisfy $x_1^2 + x_2^2 = 1$) and thus form a sphere. If we consider the above situation topologically, as discussed in Chapter 1, then we can imagine these two hemispheres to be made of two square rubber sheets. These two sheets are glued together at the edges as illustrated in Figure 3.1. Note that the direction in which they are to be glued is such that the arrows "match". We see that arrows drawn on the edges of two squares go around in different directions (see Figure 3.1).

Next, let us look at equation (2). Similarly to the previous equation, for the three-dimensional sphere S^3 we have

the set of points that satisfy equation (2) and have $x_4 \geq 0$, represents the "northern hemisphere" of S^3

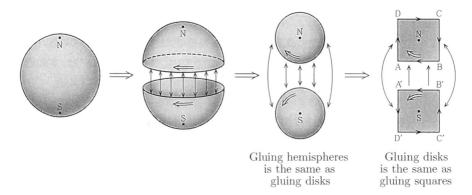

Gluing hemispheres
is the same as
gluing disks

Gluing disks
is the same as
gluing squares

FIGURE 3.1

and

the set of points that satisfy equation (2) and have $x_4 \leq 0$, represents the "southern hemisphere" of S^3.

The "equator" of the three-dimensional sphere S^3 is the set of points where $x_4 = 0$. Therefore, it is the set of all points which satisfy the equation $x_1^2 + x_2^2 + x_3^2 = 1$. This means that the equator of S^3 is a two-dimensional sphere S^2. Although we cannot directly illustrate the gluing of the two hemispheres, from the topological point of view this can be done. Since it is possible to think about the sphere as the surface (boundary) of a solid ball, we can think of the equator S^2, from this topological point of view, as the surface of a cube, as discussed in Chapter 1. In the same say as we obtained the sphere S^2 by gluing two squares, we can now obtain the three-dimensional sphere S^3 by gluing two cubes D and E (that is, by stretching and gluing them at every face). The center of one cube corresponds to the "north pole", and the center of the other corresponds to the "south pole". Therefore, as we will see later, in S^3, there can be such a thing as, say, the west wind blowing on the northern hemisphere of the earth.

The gluing of cubes D and E is illustrated in Figure 3.2. Each face of each cube is glued to the corresponding face of the other cube.

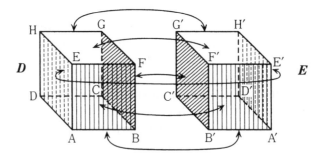

FIGURE 3.2

Now imagine cutting out of D a cylinder perpendicular to one face of D and doing the same with E using the corresponding face of E. Delete these cylinders from D and E. In the illustration, the cylinders appear to be detached from one another, but in reality in S^3 their bases are glued together. With these operations, a doughnut is carved from S^3 (see Figure 3.3).

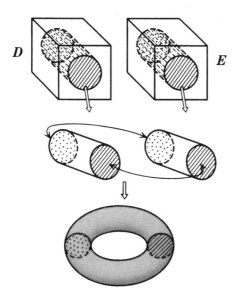

FIGURE 3.3

When the doughnut is removed, what is the rest of S^3? Let X be the rest of the cube D after the cylinder is removed, and let Y be the rest of the cube E after the cylinder is removed. Here again, we need to consider everything from the point of view of topology. This means that we must treat X and Y as being malleable and "cut-and-glue"able.

The sphere S^3 with the "doughnut" cut out is shown in Figure 3.4. Part (I) shows the cored cubes X and Y. Now imagine cutting the upper portion of X and Y and then stretching the cut parts open, as in (II). Gradually spread the two cut cored cubes open wider, as in part (III). Finally, as in part (IV), stretch the portions of the cored cubes marked with horizontal lines to the point that the portions marked with dots form the sides of "boards". Now press both sides of the "boards" and then stretch the boards in the direction of the arrows so that they form "mounds", as shown in part (V). Now glue the flat (bottom) parts of the "mound" together (note that the bases of these "mounds" are actually what once used to be the faces of the original cubes) in such a way that they together form a fat cylinder, as in part (VI). Now stretch the cylinder to elongate it, as in part (VII). Finally, bring the ends of the cylinder together and join them to form a doughnut, as in part (VIII).

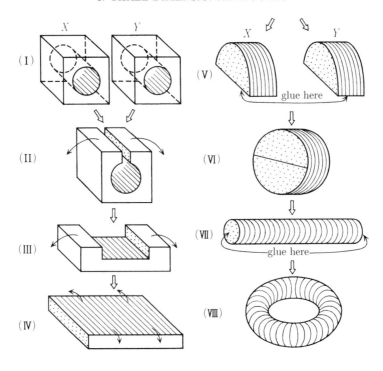

FIGURE 3.4

Thus, we see that the portion of S^3 that remains after the first doughnut was removed is itself a doughnut. To state this in a converse fashion, S^3 can be constructed by gluing the surfaces of two doughnuts together! If you pay attention to how the gluing is done, you will notice that the "rings" formed by the arrows around one doughnut are different from the "rings" around the other (see Figure 3.5), yet the two doughnuts, when glued together to

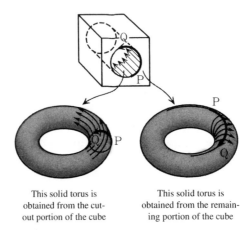

This solid torus is obtained from the cut-out portion of the cube

This solid torus is obtained from the remaining portion of the cube

FIGURE 3.5

make S^3, are glued along these rings. This may seem quite strange, but this is because we live in three-dimensional space and look at this gluing from the three-dimensional perspective. Probably, in a four-dimensional space, such gluing would seem quite natural. We will now explain this.

As we know, the torus is constructed by identifying the opposite sides of a square $ABCD$. If we identify AB and DC first, we obtain the torus (I) illustrated in Figure 3.6, and if we identify AD and BC first, then we obtain the torus (II) in Figure 3.6.

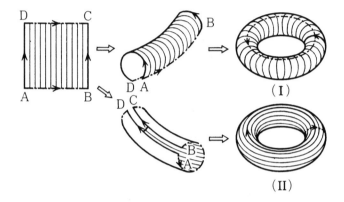

FIGURE 3.6

To reconstruct the situation where one doughnut was removed from S^3, we need to fold the sides of a square so that the face with arrows drawn on it become the *outer part* of the torus (I), and then fold the sides of the identical square in such a manner that the face with arrows drawn on it become the *inner part* of the torus (II). The solid line rings and the dotted line rings drawn on the torus in Figure 3.6 have basically the same nature. But when we try to construct these tori in three-dimensional space, they appear quite different from one another. This is a good illustration of the fact that such a figure viewed in a three-dimensional space is distorted or misperceived optically. In any case, S^3 is constructed by gluing the surfaces of the two doughnuts together. Hopefully, by now you have a better understanding of what S^3 is like. In the next section, we will provide some geometric explanations of S^3 to advance your understanding.

Now that the structure of S^3 is comprehensible, we can show that there exists a vortex-free flow on S^3, in contrast to S^2, where each flow has at least one vortex. To demonstrate this, imagine a flow of water at the angle of $45°$ in the square $ABCD$, as shown in Part (I) of Figure 3.7. Now imagine two different tori constructed from the square $ABCD$ by gluing together opposite sides. Then each stream line on each torus represents a path of particles of water on the torus. If we glue these two tori together in the manner mentioned earlier, then the trajectory lines of the water flow are also glued together. In addition, the water flow on the surface of a doughnut can be extended into the interior of the doughnut so that the water flow inside follow a circular path.

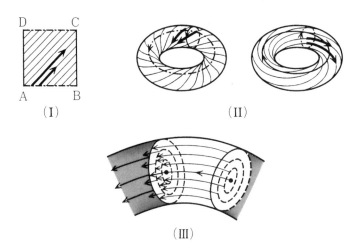

$$\textsc{Figure } 3.7$$

Therefore, we can say that the water flow inside one doughnut is glued to the water flow inside the other doughnut along the trajectory lines on the surfaces of these doughnuts. Thus, the water flow circulates continuously in S^3.

3.2. General Three-Dimensional Figures

In the previous section, you might have been puzzled by the nature of a three-dimensional sphere. Perhaps you encountered such a "figure" for the first time. Note that even though we use the word "figure", it is different from a two-dimensional surface, and it is rather difficult to visualize. In this section, we study the three-dimensional sphere in more detail. Then, we will look at how general three-dimensional figures are constructed ("figures" discussed here are smooth, just as two-dimensional surfaces). It is interesting to note that although we can explore the construction of three-dimensional figures, we cannot yet solve the problem of their classification. The classification of three-dimensional figures remains one of the most important unsolved problems in mathematics. In our discussion, we also will briefly examine the current efforts of mathematicians to solve this problem.

♣ In this discussion, a three-dimensional figure is referred to as a **three-dimensional manifold**.

A simple way to discuss a three-dimensional figure is to study a two-dimensional figure (a two-dimensional surface) first. We studied two-dimensional surfaces in Chapter 1. Now we need to add one more fact. Recall that in Chapter 1 we discussed a g-man life saver as a surface in three-dimensional space. We called it a closed surface of genus g, and denoted it by S_g. There could be many different shapes and sizes of g-man life savers, but we will ignored such discrepancies in size and shape, and looked at the figure from the topological point of view. Any surface of genus g was called S_g. Conversely, g-man life savers of various shapes and sizes can be realized in our three-dimensional space starting from just one surface S_g. According to the Gauss–Bonnet theorem, the total curvature of the surface does not depend on how S_g is realized in the space, and is equal to the Euler characteristic $\chi(S_g)$ of S_g, which is equal to $2 - 2g$ ($\chi(S_g) = 2 - 2g$), multiplied by 2π.

Let us now pay closer attention to the sign of the total curvature. We can see that if $g = 0$ (S_g is a sphere), then the total curvature of S_g is positive. If $g = 1$ (S_g is a torus), then the total curvature of S_g is 0. If $g \geq 2$ (S_g is a g-man life saver), then the total curvature of S_g is negative. Consider an infinite sequence of closed surfaces S_g, where $g = 0, 1, 2, \dots$, and divide them into three classes corresponding to positive, zero, and negative curvatures as follows:

> a sphere,
> a torus,
> a closed surface of genus $g \geq 2$.

As we said at the end of Chapter 1, each of these closed surfaces can be realized in a space of sufficiently high dimension as a surface of constant curvature 1, 0, and -1, respectively. This fact can be expressed as follows:

> There are three different two-dimensional geometries.

Now we ask: What classes of three-dimensional figures do exist and how many three-dimensional geometries are there?

First, consider the construction of three-dimensional figures from the material we already have. We will use a two-dimensional figure, that is, a closed surface S_g, and a one-dimensional figure (a circle) (see Figure 3.8).

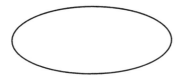

FIGURE 3.8

Let us introduce a new construction. If we have two figures K and L, we can construct a third figure called their **direct product**. We denote the direct product by $K \times L$. For simplicity, we will illustrate the notion of direct product by giving a simple example rather than a precise definition. First,

> line segment $\times$ line segment $=$ square   (see Figure 3.9).

Part (I) of Figure 3.9 makes it clear.

Next, what is the following product: line segment $\times$ circle?

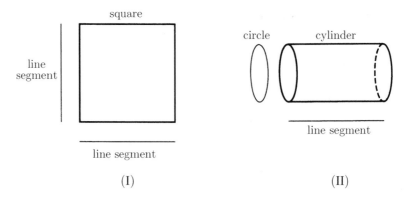

(I)   (II)

FIGURE 3.9

As you no doubt have already realized, the result is a cylinder (Part (II) in Figure 3.9). So,

$$\text{line segment} \times \text{circle} = \text{cylinder}.$$

In general, the direct product of two figures K and L is the set of all ordered pairs (P, Q), where P runs through the figure K and Q independently runs through the figure L. It is denoted by $K \times L$. For example, if we express a number line as $\mathbb{R}$, then $\mathbb{R} \times \mathbb{R}$ is a plane, that is,

$$\mathbb{R} \times \mathbb{R} = \mathbb{R}^2 \quad \text{(a plane)}.$$

Another example is

$$\mathbb{R}^2 \times \mathbb{R} = \mathbb{R}^3 \quad \text{(three-dimensional space)}.$$

In general,

$$\mathbb{R}^m \times \mathbb{R}^n = \mathbb{R}^{m+n}.$$

The direct product of an m-dimensional figure and an n-dimensional figure is an $(m + n)$-dimensional figure.

How what is the figure constructed from two circles, that is, the direct product circle $\times$ circle? The resulting new figure is a closed two-dimensional surface, hence, it is one of closed surfaces S_g. As you can see in Figure 3.10, it is a torus.

As you know, we have divided two-dimensional surfaces into three different classes. The torus belongs to one of these classes. We see that a torus can be expressed as a direct product: one-dimensional figure $\times$ one-dimensional figure, or, in other words, a torus can be decomposed into two one-dimensional figures.

With this understanding, let us now consider the direct product of a one-dimensional figure and a two-dimensional figure. The resulting figure will be three-dimensional. In general, the change of order in the product does not change the product, i.e., $K \times L = L \times K$. Hence, from circles

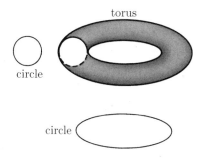

FIGURE 3.10

and three different surfaces we can construct new figures using the direct product:

>   circle × sphere
>   circle × torus
>   circle × closed surface with genus g.

The resulting figures are three-dimensional. They are constructed as three-dimensional figures in which a closed surface is placed at each point on a circle. Study Figure 3.11 and use your imagination to visualize the three-dimensional figure in the five-dimensional space. Since we live in three-dimensional space, the figure intersects itself (as in the case of a Klein bottle). Since our imagination can "leap" beyond our dimension, we can simply undo self-intersections in our minds.

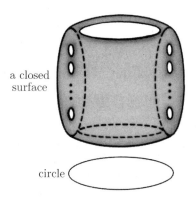

FIGURE 3.11

Can we construct new three-dimensional figures from the material we have at hand (circles and closed surfaces) without using direct product? To understand the construction of three-dimensional figures, we return to the study of two-dimensional figures.

We already know that a torus is the direct product of two circles. That is, a torus = circle × circle. Expressed in a different manner, a torus is a two-dimensional figure where a circle is placed at each point of another circle, and the these circles sitting on another circle are bundled together. Let us now construct another figure by placing circles on another circle as described above, but bundle them differently. First, since we obtain a circle by gluing together the endpoints of a segment, we will construct the direct product of a segment and a circle (segment × circle). The resulting surface is a cylinder. If we glue the circles at both ends of the cylinder (top base and bottom base), we simply obtain a torus, as discussed in Section 3.1. However, if, instead, we glue the two bases together, with the direction of the circles reversed, then the resulting figure is a Klein bottle.

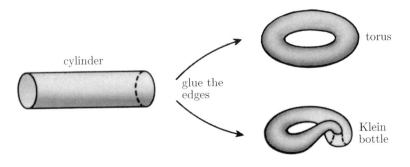

FIGURE 3.12

Thus, we see that a Klein bottle can be obtained by placing a circle at each point of another circle in the same manner as in the case of a torus. However, a Klein bottle is constructed by twisting the bundle of infinitely many circles. We express this process as follows:

$$\text{Klein bottle} = \text{circle} \mathbin{\widetilde{\times}} \text{circle}$$

("$\widetilde{\times}$" denotes the **torsion product**, whereas "$\times$" denotes the direct product).

We can generalize the notion of the torsion product as follows.

If there are two figures K and L, then the torsion product $K \mathbin{\widetilde{\times}} L$ is the figure obtained by placing the figure L at each point of the figure K (as in the case of a direct product) and then twisting the resulting "bundle" of infinitely many duplicates of K. The output of this manipulation is the torsion product $K \mathbin{\widetilde{\times}} L$. Thus, one obtains different torsion products depending on the order of figures K and L. In general, the figure $K \mathbin{\widetilde{\times}} L$ and the figure $L \mathbin{\widetilde{\times}} K$ are different.

Using the notion of torsion product, we can construct many new three-dimensional figures from the materials at hand (circles and closed surfaces). There are six different torsion products which can be obtained as various combinations of circles and surfaces. These are

sphere $\widetilde{\times}$ circle

torus $\widetilde{\times}$ circle

closed surface of genus $g \geq 2 \mathbin{\widetilde{\times}}$ circle

circle $\overset{\sim}{\times}$ sphere

circle $\overset{\sim}{\times}$ torus

circle $\overset{\sim}{\times}$ closed surface of genus $g \geq 2$.

By adding the three direct products mentioned earlier to these torsion products, we obtain eight different constructions (instead of nine). This is because one of the torsion products, circle $\overset{\sim}{\times}$ sphere, is nothing but a direct product, circle $\times$ sphere (as long as we consider only orientable surfaces).

An American mathematician William Thurston, who is at present a professor at Cornell University, has produced several significant theories pertaining to geometric figures, and has contributed greatly to the advancement of modern geometry. One of the theories he developed is related to three-dimensional figures. When his theory is completed, we will know a great deal more about the fundamentals of three-dimensional figures. The substance of the theory is summarized as follows.

Any three-dimensional figure can be obtained by combining the previously mentioned eight constructions. The construction of each figure is unique.

This is called "Thurston's Geometrization Conjecture" or "Thurston's Program". There are many mathematicians in the world working to develop this conjecture further even as you read this book.

So far we were discussing rather abstract topics. Let us now return to more concrete subjects, such as the three-dimensional sphere. Since we all know that the three-dimensional sphere is a three-dimensional figure, if Thurston's program is correct, we should be able to construct the three-dimensional sphere using one of eight constructions or a combination of two or more of them. In fact, the three-dimensional sphere has the structure of the torsion product sphere $\overset{\sim}{\times}$ circle. Expressed in a different manner, the three-dimensional sphere is a figure obtained by placing a circle at each point of a sphere, then bundling the totality of circles in a particular fashion. Conversely, we can completely fill the interior of the three-dimensional sphere with an infinite number of circles without allowing the circles to intersect each other. (This astonishing fact was discovered by Heinz Hopf many years before Thurston made his conjecture—in fact, even before Thurston was born.)

This fact is explained by the existence of a circular flow on a three-dimensional sphere (as discussed in Section 3.1). To understand the above phenomena, let us examine the three-dimensional sphere from a more general and geometric point of view. This point of view often requires employing some special devices. One of these devices is called the **point at infinity**. To understand the concept of the point at infinity, it is necessary to use your imagination.

First, let us explain the concept of the point at infinity. When you studied the number line in high school, you were instructed to use the notation $+\infty$ if the point moves to the right of the origin without a bound, and $-\infty$

if the point moves to the left of the origin endlessly. However, your instructor probably told you that there is no actual point called "infinity". But mathematicians nonetheless use the notion of the point at infinity. That this is a necessary tool of the trade is revealed in simple situations such as when a rubber band (a circle) is cut at a given point to become a string (a line)! Similarly, the ends of the string can be glued back together with a single drop of glue to form the band (circle) once again. Such real-life facts demonstrate the abstract principle that one obtains a circle by adding one point to a line and gluing the "infinite" ends together (see Figure 3.13). The point attached to the line is called the point at infinity. Simply put,

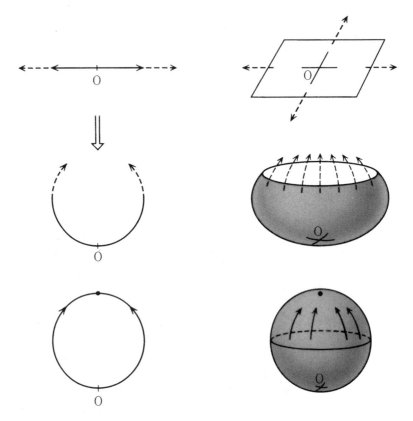

FIGURE 3.13. In dimension one, a line becomes a circle by adding the point at infinity. In dimension two, a plane becomes a sphere by adding the point at infinity.

$$\text{circle} = \text{line} + \text{the point at infinity}.$$

We claimed earlier that a circle is a one-dimensional figure. But let us look at a two-dimensional figure and discuss the point at infinity on the two-dimensional plane. If we think of the surface of the Earth as a rubber ball, and imagine "puncturing" it at the north pole, then topologically, we can flatten the remaining surface. In other words, we can obtain a sphere by adding to a bent plane a point corresponding to the "north pole" (see Figure 3.13). Maybe you can imagine this situation as wrapping a soccer ball with a handkerchief with its corners tied together. The point at which the corners of the handkerchief meet is the point at infinity. That is,

$$\text{sphere} = \text{plane} + \text{the point at infinity}.$$

We will now extend this analogy into three-dimensional space, that is, we imagine three-dimensional space spreading out infinitely far along any direction. Let us assume that infinite "edges" eventually meet at a single point. We call this imaginary point the **point at infinity**. Adding the point at infinity to three-dimensional space, we obtain a three-dimensional sphere. That is,

three-dimensional sphere = three-dimensional space + the point at infinity.

This will be at the center of the following discussion. If you can visualize a three-dimensional sphere as three-dimensional space together with the point at infinity, the following discussion will be easier to understand.

Two circles embedded in $\mathbb{R}^3$ as in Figure 3.14 are called the Hopf link. If we consider these two circles in a three-dimensional sphere, one of the

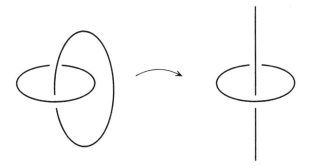

FIGURE 3.14

circles can be made very large making it so large that it passed through the point at infinity. When that happens, the circle will look as a line of infinite length. In fact, both "ends" of the line are glued together at the point at infinity, resulting in a circle. A pair of circles, one of which passes through the point at infinity, is also called the Hopf link (in the three-dimensional sphere). In Figure 3.14, the vertical line passes through the point at infinity, and this line can be regarded as a circle if the point at infinity is added.

We now have two circles linked together and embedded in a three-dimensional sphere. We will show that these two circles can be joined by an infinite number of intermediate circles, and that the three-dimensional sphere is filled with these circles. The best way to see this is to add material to the "core" circle C (not the vertical line) in Figure 3.15. Consider the end result to be the doughnut D in Figure 3.15, and the surface of the doughnut is, as you know, a torus. The circle C is the core of the doughnut.

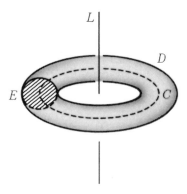

FIGURE 3.15

If you slice vertically through the left side of the doughnut D, the resulting cross-section will be a disk E as in Figure 3.15. Since the center of E is on the circle C, if you move E along the circle C, then the disk E will sweep the entire doughnut D. In other words, the doughnut D is the direct product of the disk E and the "core" circle C. That is,

$$\text{doughnut } D = \text{disk } E \times \text{circle } C.$$

What would remain if we removed the doughnut D from the three-dimensional sphere? Recalling the discussion in the previous section, one might guess that it is another doughnut. This is correct, but let us now explain in a specific way how the remaining figure becomes a doughnut. There is a vertical line (which, if joined at the point at infinity, becomes a circle) through the remaining figure. We call this line L. Now consider the plane T in which the circle C lies. Place a disk E' on the plane T so that the center of the disk E' be at the intersection point of the line L and the plane T, as in Figure 3.16. The disk E' now looks as a "membrane" covering what would otherwise be the hole in the doughnut. Now move the disk E' up and down along the line L keeping the boundary of the disk in contact with the surface of the doughnut at all times. This manipulation should yield figures resembling soup bowls (an upside-down one on the "positive" side of the vertical axis formed by line L, and an ordinary soup bowl on the "negative" side). If we continue to enlarge the "soup bowls", the figure should begin to look like a hot air balloon and eventually will "swallow" all points outside of the doughnut D. If this seems confusing, refer to Figure 3.16.

What shape does the figure (which began as the disk E') have when it reaches the point at infinity? If we look at this figure in $\mathbb{R}^3$ (viewing $\mathbb{R}^3$ as a three-dimensional sphere without the point of infinity), the figure will have a hole in the center (the center of the figure is the point at infinity) and the region surrounding the hole spreads to infinity in all directions. In Figure 3.17, you can see that the figure is the shaded region outside of the intersection of the doughnut and the plane T where the core disk E' lies, excluding the hole of the doughnut.

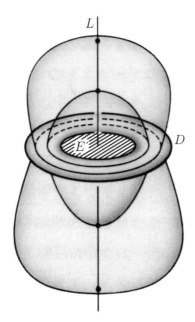

FIGURE 3.16

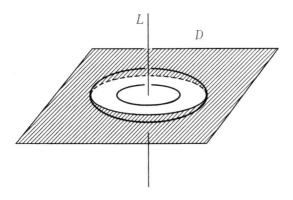

FIGURE 3.17

To summarize the above discussion, suppose $\bar{L}$ is the circle in the three-dimensional sphere obtained by adding the point at infinity to the line L. When the disk E' revolves along the circle $\bar{L}$ as described above, it sweeps the region of the three-dimensional sphere with the doughnut removed. In other words,

$$\text{three-dimensional sphere} - \text{doughnut } D = \text{disk } E' \times \text{circle } \bar{L}.$$

Looking at the right side of this equation, you will notice that it is the direct product of a disk and a circle, and that is a doughnut. Let us call this doughnut D'. Then

$$\text{three-dimensional sphere} = \text{doughnut } D + \text{doughnut } D'.$$

Hence, the three-dimensional sphere is made of two doughnuts.

You may recall that this fact was demonstrated in the previous section, though from a different point of view. Here, using concrete figures, we have explained how a three-dimensional sphere is constructed in our three-dimensional space.

> ♣ The doughnuts D and D', as constructed above, corre-spond to the two doughnuts in Section 3.1. Yet, you might have noticed that the doughnuts mentioned in Section 3.1 look the same, whereas the doughnuts D and D' above do not look the same. The reason that D and D' do not look alike is because we draw these figures by regarding the point at infinity as a special point of the three-dimensional sphere. If we observe D and D' in the situation where no point of the three-dimensional sphere is of more importance than any other point, then these two doughnuts should look the same.

Let us now construct the flow of fluid in a three-dimensional sphere as discussed in Section 3.1. The three-dimensional sphere is described as the sum of two doughnuts D and D', where D has the circle C as its "core" circle, and D' has the circle $\bar{L}$ as its "core" circle. By enlarging or shrinking the doughnut around the "core" circle we can obtain doughnuts of various shapes. By combining these two facts, we can describe the three-dimensional sphere as follows.

First, look at the figures in Figure 3.18. They show how one constructs and enlarges a doughnut starting from the "core" circle C. We obtain a doughnut D by adding material to (or "thickening") the "core" circle C. After the doughnut D is built, we continue to add more material "layer by layer" around D, thereby "thickening" it. Eventually, D will enclose the circle $\bar{L}$, and it will cover the whole space with the exception of the circle $\bar{L}$. Then, if we add the circle $\bar{L}$ to the doughnut at the end of all enlargements, the resulting figure will be the three-dimensional sphere.

We are now ready to construct a model of the flow of fluid by placing infinitely many nonintersecting circles in the three-dimensional sphere. For the time being, we will leave the circles C and $\bar{L}$ as they are. We know from the previous discussion that there are infinitely many "layers" of doughnut surfaces (like layers of Filo® dough) between the "core" circles C and $\bar{L}$. Each of these layers looks like a torus. Imagine drawing circles on each layer of the surface of the doughnuts (as in Figure 3.19). Each circle goes around the surface of the doughnut once, passing through the "hole" and coming back from the other side; thus, it rotates not only around the core but also in the direction of any given cross-section of the doughnut. (See illustrations (a) and (b) in Figure 3.19 for clarification.)

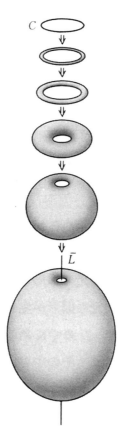

FIGURE 3.18

To construct a flow of a fluid in a three-dimensional sphere we have drawn only eight circles (Figure 3.20). In fact, there is an infinite number of

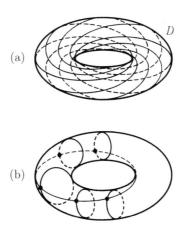

FIGURE 3.19

circles filling each surface, or, to put it differently, each surface is decorated with an infinite pattern of circles. By drawing this pattern of circles (which represents the flow of fluid) on each surface, we see that even as we come closer and closer to the circle $\bar{L}$ and examine the vertical patterns, the circles are glued together neatly without disrupting the patterns. This illustrates the circular flow described by Hopf, and it shows that the three-dimensional sphere is filled with an infinite number of circles.

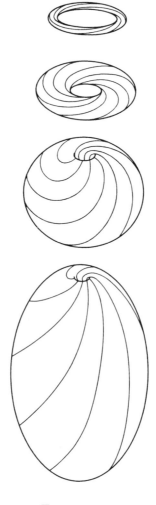

FIGURE 3.20

CHALLENGE 6. Take any two circles from the set of circles constructed in the three-dimensional sphere. They will become a Hopf's link after a continuous deformation. In other words, any two circles constructed in the three-dimensional sphere are linked together. Explain this by drawing appropriate figures.

Next, consider what kind of figures we would obtain if each circle in the three-dimensional sphere were reduced to a point. Intuitively, we see that the resulting figure is two-dimensional since one-dimensional circles are reduced to a point in a three-dimensional figure. Indeed, each circle drawn on each "layer" surface in the doughnut D intersects the disk E once. Therefore, if we reduce each circle to a point, the resulting figure is the two-dimensional disk E. Similarly, each circle drawn on each "layer" surface in the doughnut D' intersects the disk E' once (a torus is hollow, while a doughnut is solid).

To put this differently, we can say that each circle inside the doughnut D starts from and returns to a point in the disk E. Also, each circle inside the doughnut D' starts from and returns to a point in the disk E'. In this manner, we can associate circles in the three-dimensional sphere with points in disks as follows:

a circle inside the doughnut $D \rightarrow$ an intersection point in E;

a circle inside the doughnut $D' \rightarrow$ an intersection point in E'.

To understand the relationship between circles and points, imagine infinite circles "originating" and "closing" at each point in E and E' (see Figure 3.21) so that the totality of such circles fills the entire three-dimensional sphere. These circles generally neither overlap nor intersect except for the case where the circles are on the surface of the corresponding doughnuts D and D'. In these exceptional cases, the circles overlap, or, more precisely, they become one circle performing double duty. If we look at a particular circle C from the D perspective, that is, if we take it to be on the surface of D, then circle C can be said to "originate" from a point P in E. However, that same circle also lies on the surface of D'. Therefore, we can also say that it "originates" at a point Q in E' (see Figure 3.21).

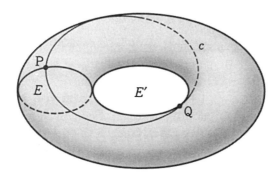

FIGURE 3.21

Only one circle will pass through the points P and Q (which are on the boundaries of E and E', respectively). Circles originating at interior points of E and E' are unique. If we regard the two disks E and E' as the result of reducing circles in D and D', we need to glue the boundaries of E and E'

together. As a result of the gluing, we obtain the two-dimensional sphere (see Figure 3.22).

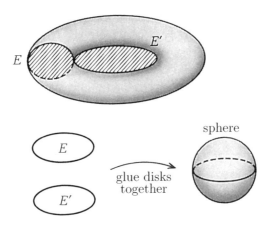

FIGURE 3.22

More precisely, we obtain the three-dimensional sphere by arranging a circle at each point of the two-dimensional sphere and then bundling the totality of circles in a certain manner, i.e.,

$$\text{three-dimensional sphere} = \text{two-dimensional sphere} \overset{\sim}{\times} \text{circle}.$$

The three-dimensional sphere is the torsion product of a two-dimensional sphere and a circle!

Let us now go back to the discussion of the eight different constructions of three-dimensional figures. These constructions are divided roughly into three classes: the direct product of a surface and a circle, the torsion product of a surface and a circle, and the torsion product of a circle and a surface. The construction of the three-dimensional sphere belongs to the second class.

Precisely speaking, it is the torsion product of a sphere and a circle. This class of construction and its generalization have been studied since the 1930s, and many results have been published. Together, these results constitute an almost complete theory. Today, intense research focuses on the third class, the construction by the torsion product of a circle and a surface. Researchers in this field study in detail one-to-one correspondences between surfaces (which have an astonishingly rich structure if $g \geq 2$).

Have you heard the term "knot"? A knot is an embedding of a circle in $\mathbb{R}^3$ (see Figure 3.23).

By embedding the circle in various ways, we can create many different knots in $\mathbb{R}^3$. If we remove a knot from $\mathbb{R}^3$ and put a cover on the "hollow" tube in $\mathbb{R}^3$ and the point at infinity in a particular fashion, we obtain a closed three-dimensional figure. By removing various "knots" from $\mathbb{R}^3$ and covering each of the resulting "hollow tubes" and the point at infinity, we obtain various closed three-dimensional figures. One of yet unsolved

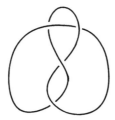

FIGURE 3.23

questions about the representation of three-dimensional figures as torsion products is whether or not these figures can be obtained by applying simple manipulations to the torsion product of a circle and a sphere.

This concludes our discussion of three-dimensional figures. We hope you have enjoyed your journey through the rich and diverse landscape of three-dimensional space!

Since we previously discussed three-dimensional figures, we can now think about four-dimensional, five-dimensional, six-dimensional, and in general, n-dimensional figures. In fact, modern geometry can deal with these higher-dimensional figures as if they could be visualized, and this sheds light on many interesting phenomena. One might assume that the higher the dimension, the more elaborate the figures we construct and the more difficult the problems or questions we answer. There is some truth to this; in some sense, higher-dimensional figures are more complicated. But recent developments in modern geometry show that in the three-dimensional and four-dimensional space, we encounter difficulties characteristic to these spaces, which disappear if the space has a higher dimension! There are numerous questions about the three-dimensional and four-dimensional space, especially the latter, which is still shrouded in mystery. In recent years, it has become apparent that the four-dimensional world can best be understood by a combination of disciplines, particularly geometry and physics. With the help of physics, many astonishing facts have been proven. Yet this has served only to deepen the mystery of the four-dimensional world.

It seems as if there were a special meaning attached to the three-dimensional and four-dimensional space. We do not know whether this special meaning is merely a present-day phenomenon or whether it represents an eternal philosophical truth. In any case, for the foreseeable future, three-dimensional and four-dimensional geometry (and their interrelationship) will be a popular subject. We encourage our readers to accept the challenge to tackle these problems themselves.

Physics and Dimension

A.1. Newtonian Mechanics

In the ancient world, people experienced the passage of time by watching certain natural phenomena such as the alternation of day and night, the waxing and waning of the moon in the night sky, and the change of the seasons. While observing the immensity of the blue sky during the day and the dark sky full of sparking stars at night, they must have perceived that spread above the ground upon which they stood, there existed a vast space. It seems reasonable to believe that at some point, they developed a vague understanding of time and space.

In ancient civilizations, geometry and arithmetic were born out of the necessity to survey the land in order to assess taxes and to construct sturdier and more complex buildings. But by the time of ancient Greeks (some twenty four hundred years ago), geometry became the study of properties of figures themselves, not simply a tool to aid in surveying the land or constructing buildings. This became the foundation of present-day mathematics. Euclid was responsible for synthesizing this geometry into the remarkable work called *Elements* in approximately 300 B.C. All the achievements of Greek mathematicians are encompassed in his work consisting of thirteen volumes (not all of which have survived). His work contained the theory of **plane geometry**, which deals with plane figures such as triangles and circles, and **solid geometry** which deals with solid figures, such as polyhedra. Euclid's work established the foundation of geometry. The ancient Greeks believed that the space they live in is literally the world described in *Elements*, that is, the world with Euclidean geometry. They took Euclid's word for gospel, even describing the movement of planets in terms of Euclidean geometry. The astronomical work, advanced, in particular, by Claudius Ptolemaeus (approximately 150 A.D.), was very precise. His perception of the world persisted throughout the middle ages, until Copernicus claimed that the Earth and the planets revolved around the sun.

Solid geometry deals with solid figures which have length, width, and height. Therefore, we can say that it is the geometry of the three-dimensional world. This geometry is compatible with the world we live in: the surface of the Earth has length and width, and the universe above us gives us height.

It has long been taken for granted that the world that surrounds us is three-dimensional and that time passes through this world uniformly from the past to the future.

This geometric view of our world was described by ancient Greek mathematicians as **properties of figures**. They arrived at this view by employing the **axiomatic approach**. That is to say, their geometry was developed from a handful of axioms. Much later, in the 17th century, René Descartes and Pierre de Fermat introduced the coordinate system. Thereafter, mathematicians could describe geometry and the motion of objects by using coordinate systems and equations. Still later, Isaac Newton (1643–1727) established the study of calculus with the help of classical mechanics. Owing to these great achievements, further development of mathematics and science has been exceedingly rapid.

In Newtonian mechanics, the motion of an object (a point mass) is described by differential equations. If we know the position and the velocity of an object at a certain time, we can completely understand the motion of the object. For example, the motion of an object upon which no force is exerted, starting from the origin, is described by the following equations:

$$x = at, \quad y = bt, \quad z = ct,$$

where we introduced an orthogonal coordinate system (x, y, z) in the space (see Figure A.1). The above equations describe the motion of an object with a constant velocity. The variable (t) in the above equations represents time. It describes the movement of an object as a parameter, independent of the coordinates (x, y, z) in space. As you can see, the significance of Newtonian mechanics is that it assumes, as a condition of understanding the physical world, that the three-dimensional space (where the object moves) and time are absolute. Time is considered to be independent of space.

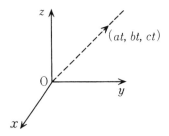

FIGURE A.1

Although the person generally credited for the discovery of calculus is Newton, Gottfried Wilhelm Leibniz (1646–1716), who was a mathematician and philosopher around the same time as Newton, is believed to have established the study of calculus independently of Newton. However, Leibniz' understanding of the physical world was far different from Newton's. Leibniz criticized Newton's idea of absolute space and absolute time, and he found Newton's idea impossible to accept. However, the results of subsequent experiments and astronomical measurements supported the accuracy of Newtonian mechanics and Leibniz' criticism was eventually forgotten.

A.2. Geometry Describing Space

The significance of Euclidean geometry is found in the Fifth Postulate in *Elements*, called the "Parallel Postulate". The Parallel Postulate is usually expressed as follows: the necessary condition for two lines to be parallel is that the alternate interior angles are congruent (see Figure A.2). According to the definition, two lines l and m are parallel if they do not intersect. We do not know whether two lines are parallel unless we follow the lines towards infinity and examine whether they intersect at infinity. However, according to the Parallel Postulate, we can determine if two lines are parallel simply by checking the alternate interior angles which can be observed (not necessarily at infinity). This postulate allows us to determine something which otherwise would be a practical impossibility; therein lies its great strength and, some would say, mystery. This also reminds us of the **Law of Inertia** in Newtonian mechanics, which says that a moving object is in motion with a constant velocity unless a force is exerted upon it.

FIGURE A.2

It is known that the Parallel Postulate is equivalent to the statement that the sum of the interior angles of a triangle is 180°. This equivalent result gradually became clear through the process of trial and error as mathematicians attempted to deduce the Fifth Postulate from other axioms. Giovanni Saccheri (1667–1733) and Johann Heinrich Lambert (1728–1777) were two such mathematicians. They showed that if the Fifth Postulate (the Parallel Postulate) does not hold, then the sum of the interior angles of a triangle is either greater or less than 180°.

János Bolyai (1802–1860) and Nikolai Ivanovich Lobachevsky (1793–1856) discovered, independently of each other, a geometry different from Euclidean geometry, in which the sum of the interior angles of a triangle is less than 180°. This non-Euclidean geometry is nowadays called **Hyperbolic Geometry**. A geometry in which the sum of interior angles of a triangle is *greather* than 180° is called **Spherical Geometry**, and is the geometry of a sphere. It appears that Gauss, too, discovered early on that there was a possibility of the existence of a non-Euclidean geometry. However, he was reluctant to publish his discovery. In Gauss' time, the concepts of Euclidean geometry were closely tied to the basic ideas of the natural philosophy, and Gauss had reason to fear public criticism or worse. However, he did conduct experiments measuring the sum of the angles in a triangle to determine if the geometry of the physical world was, in fact, Euclidean (as long had been believed). Gauss performed this experiment by picking three widely separated points (three mountain peaks) as the vertices of a gigantic triangle. He then measured the sum of angles of the resulting triangle. The results of the experiment, however, were inconclusive because of large measurement errors in the experiment. What Gauss' experiment does tell us is that if we measure distances that are not too large (ordinary distances we encounter in our daily life), Euclidean geometry is useful. It accurately represents the physical world in which we live, on a smaller scale.

On the other hand, Riemann, who was greatly influenced by Gauss, pondered on even more general geometry (see the sixth week of "A Story of the Development of Mathematics," authored by Shiga, published by Iwanami Shoten). Riemann asserted that geometries were not given to man by Heaven, but resulted from our conceptual view of the world as the manifestation of our recognition of space. To describe these abstract geometries, he introduced the notion of an n-dimensional manifold, and today we know these geometries as Riemannian Geometry. Riemann described geometries from an entirely new perspective. Although Riemann seemed to believe that our world is three-dimensional, his conception of geometries aimed towards higher dimensions. Furthermore, it appears that he also considered the possibility of a discrete space (a world which consists of an enormous number of particles).

A.3. From Newtonian Mechanics to Relativity

At the end of the nineteenth century, A. A. Michelson and E. W. Morley carried out experiments designed to determine the existence of "ether", a medium through which light was assumed to propagate. During their experiments, they discovered a misterious fact related to the velocity of light. Their experiments produced evidence that the velocity of light in empty space is always the same, no matter what the velocity of the observers. Suppose a person at rest measures the velocity of light at a certain location and obtains the result c. Another person traveling in a train moving at a constant velocity v with respect to the person at rest also measures the same velocity c of the same light. No matter what the velocity of the observer in the train, the velocity of light is still c instead of $c + v$. The velocity v of the moving observer is not added to the velocity c of light. This discovery revealed that at high velocities (velocities near the velocity of light), Newtonian mechanics is no longer valid. A new physics was thus needed to replace Newtonian mechanics.

Around the same time as Michelson and Morley were carrying out their experiments, H. A. Lorentz made an interesting observation of events during an experiment involving two observers in constant motion. To understand Lorentz' observations, let (x', y', z') be a coordinate system moving at a constant velocity v in the x-direction with respect to the coordinate system (x, y, z). Let t be time measured by a clock in the coordinate system (x, y, z), and t' time measured by a clock in the coordinate system (x', y', z'). Lorentz concluded that there should be the following relationship between the space and time coordinates assigned to the same event observed in the two coordinate systems:

$$ t' = \frac{t - \left(\frac{v}{c^2}\right) x}{\sqrt{1 - \left(\frac{v}{c}\right)^2}}, \quad x' = \frac{x - vt}{\sqrt{1 - \left(\frac{v}{c}\right)^2}}, \quad y' = y, \quad z' = z. $$

This system of equations is called the **Lorentz Transformation**. If you look at the first equation, which is a relation between times t and t' measured in two different coordinate systems, you will notice that it contains the variable x. This shows that time is not independent of space. Therefore, the basic principle of Newtonian mechanics, which states that time is absolute and is independent of space, is no longer valid. In Newtonian mechanics, $t = t'$. This shows that if the velocity v is extremely small compared to the velocity of light, then the approximation of v/c is 0. By taking $v/c = 0$, we will obtain equations of Newtonian mechanics from the Lorentz transformation.

Einstein established the revolutionary **Special Relativity** by deriving the Lorentz transformation from the principle that the velocity of light c is constant, the fact that was verified by Michelson and Morley. In this new theory, the notion of absolute time is no longer valid; rather, space and time can mix. As a replacement for the notion of absolute time, Einstein introduced four-dimensional space-time (x, y, z, t) to represent the physical world. Hence a physical phenomenon, which in Newtonian mechanics was described by three-dimensional space coordinates (x, y, z) together with independent time t, is now described by four-dimensional space coordinate (x, y, z, t) in the special relativity. But Einstein was not satisfied by special relativity alone. Instead, he further pursued a new theory of gravitation. It involved a generalization of the principles of relativity. That new theory is **General Relativity**. The two fundamental principles of general relativity are as follows.

(1) The laws of physics in any coordinate system can be described in the same mathematical form (invariance under general coordinate transformation).

(2) The gravitational mass and the inertial mass are equivalent.

General relativity is a theory of gravity, and it can be considered a refinement of Newton's theory of gravity. General relativity was instrumental in the development of methods used to describe the physical phenomena of gravitation in the four-dimensional space-time using the geometrical framework.

After General Relativity, Einstein attempted to develop another theory. He tried to unify the **Theory of Electromagnetism** and the **Theory of Gravity** into the **Unified Field Theory**. Though he was not successful in this effort in his lifetime, his work has since been transformed, and it has resurfaced as **Superstring Theory**.

A.4. Quantum Mechanics and Quantum Field Theory

At the end of the nineteenth century, Max Planck (1858–1947) assumed that energy distribution is not continuous, but rather that energy consists of a finite number of discrete packets. This assumption is often called "Planck's quantum hypothesis", although Planck himself did not understand fully the true meaning of the granular nature of energy. Planck considered his quantum hypothesis to be a mathematical device, but Einstein recognized it as an important innovation in physics. By expanding upon Planck's work, Einstein proposed the **Photon Theory of Light**, which states, in essence, that light is transmitted as tiny particles (photons). Since then, Einstein's photon theory of light became linked to the study of quantum mechanics, which in turn had moved into the mainstream of the twentieth century physics.

After Planck's quantum hypothesis was materialized as Einstein's photon theory of light, physicists accepted the concept of **Wave-Particie Duality of Light**, which says that light exhibits both particle-like and wave-like characteristics. As the structure of atoms was better understood, it became apparent that Newtonian mechanics could no longer be applied to the microscopic world. Elementary particles, which are the building blocks of atoms, also possess this wave-particle duality. The part of physics that better describes the microscopic world of elementary particles is called **Quantum Mechanics**. Quantum mechanics was developed first by Erwin Schrödinger (1887–1961) and Werner Heisenberg (1901–1976) in the middle of the 1920s. The theoretical approaches by these two physicists were as distinct as the particles and waves they observed. Yet, even though they took very different approaches, the answers to specific questions were identical regardless of which theory was used. Bewildered physicists struggled to determine which theory was accurate. John von Neumann laid this confusion to rest when he showed that these two theories were supported by the abstract theory of **Hilbert Space**. Von Neumann asserted that, depending upon the realization of certain relations in various Hilbert spaces, the resulting descriptions of quantum mechanics would appear different. The same description is obtained by using either Hilbert space representation or Schrödinger's representation. Thus, von Neumann showed that the Hilbert space representation and the Schrödinger representation are equivalent. A good example of a Hilbert space is the set of all points x with infinitely many coordinates

$(x_1, x_2, \ldots, x_n, \ldots)$ such that $\sum_{n=1}^{\infty} x_n^2 < \infty$, equipped with the inner product $(x, y) = \sum_{n=1}^{\infty} x_n y_n$. This example corresponds to the quantum theory of Heisenberg.

In quantum mechanics, it is necessary to use infinite-dimensional spaces, called Hilbert spaces, to describe physical phenomena. Quantum mechanics predicts the probability of position and velocity of a moving particle. This notion differs slightly from what we may assume intuitively. However, in quantum mechanics, the space in which particles move is still four-dimensional space-time.

After the establishment of quantum mechanics, physicists endeavored to describe further the motion of elementary particles, but they encountered additional difficulties in their research. Elementary particles are unstable in the natural world, and even in a vacuum they are constantly being born and disintegrating. Thus, predicting the motion of a single elementary particle is difficult at best. Therefore, theoretically it would be necessary to consider an infinite number of particles.

In Newtonian mechanics, one describes the motion of a particle by the location (x, y, z) of the particle and the velocities in x-, y-, and z-directions. We say that such a particle has six degrees of freedom. This means that to describe the motion of two particles we need 6×2 degrees of freedom. For each description of motion of a finite number of particles, there is the corresponding finite number of degrees of freedom. However, describing the motion of elementary particles requires infinite number of degrees of freedom; therefore, the theory becomes rather complicated. The part of Physics that deals with elementary particles is called **Quantum Field Theory**. Because it involves infinity, to set it up as a mathematical structure requires dealing with a number of difficult problems. And since it is necessary to use the devices of physics, quantum field theory cannot be established as a precise mathematical structure. Thus, even today, we have yet to install quantum field theory as a precise mathematical structure.

By further expanding quantum field theory, it became possible to deal with the forces of electromagnetism and the forces acting between elementary particles (weak interaction and strong interaction) in a unified manner. We still have not built a unified theory which incorporates gravity.

A.5. Superstring Theory and Ten-Dimensional Space-Time

The study of string theory (more precisely, superstring theory) developed very rapidly in the 1980s, and in fact, became the theory that was supposed to deal with all known forces of the universe in a unified manner. In this theory, a particle is represented not by a point, but by a string, which has a one-dimensional structure. This theory once was thought of as the most promising construction of the unified field theory. However, this theory has a very strange feature: to preserve the invariance under general coordinate transformations, that is, to be able to describe physical phenomena using an arbitrary coordinate system, the space-time must be ten-dimensional. Yet, the idea of ten-dimensional space-time is in direct contradiction to the generally accepted idea that our space-time is four-dimensional. Many physicists were shaken by this theory.

Interestingly, when Einstein was trying to establish a theory which would unite the theory of electromagnetism and the theory of gravitation, he knew that if our space was considered five-dimensional, a wonderful five-dimensional theory of gravity could be established. However, there was no experimental evidence to support the idea of five dimensions. Thus, the five-dimensional theory of gravity was considered meaningless in physics. Given this, one cannot help but wonder at the validity of theories that assume our universe to be ten-dimensional!

To these objections, the experts in superstring theory respond that to show the structure of ten-dimensional space-time of superstring theory, extremely high energy is needed. Compared to this kind of high energy, the energy used in today's experiments is close to nothing. Since scientists cannot create this extremely high level of energy in laboratories, they are unable to see beyond four dimensions. To be able to see the structure of ten-dimensional space-time, we need far higher energy. Superstring theorists assert that our observation of physical phenomena reveals only four dimensions, while the remaining six dimensions are hidden. They further assert that the hidden six-dimensional space-time is closed, and that in the first order approximation it has the structure of a complex three-dimensional Calabi–Yau manifold.

These assertions have prompted some to scorn superstring theory researchers. And, indeed, there are no new fundamental results in superstring theory which can be verified by experiments. However, if we change our focus from physics to mathematics, the situation changes. Superstring theory has excited the front lines of research in modern mathematics as never before. For example, one physicist calculated the number of rational curves over a Calabi–Yau manifold and discovered that the result was identical to the one known in mathematics. These experiments have recently been verified.

This example shows that work on superstring theory continues to provide impetus for further research. This theory has revealed unexpected connections between various branches of mathematics. As mentioned above, superstring theory has a mathematically rich structure, even if we cannot yet understand it from the point of view of physics.

For now, we know that superstring theory contains certain physical truths. But if we explore deeper, might we encounter a discrete world? Physics of a discrete space-time has scarcely been studied. In a discrete world, dimension may manifest itself in a completely different form. There is no known comprehensive mathematical method to study a discrete world. However, we hope that superstring theory may provide a set of tools that can be used to unlock the mystery of this obscure discrete world.

Titles in This Series